Engineering Libraries: Building Collections and Delivering Services

Engineering Libraries: Building Collections and Delivering Services has been co-published simultaneously as *Science & Technology Libraries*, Volume 19, Numbers 3/4 2001.

Engineering Libraries: Building Collections and Delivering Services

Thomas W. Conkling
Linda R. Musser
Editors

Engineering Libraries: Building Collections and Delivering Services
has been co-published simultaneously as *Science & Technology Libraries*,
Volume 19, Numbers 3/4 2001.

Routledge
Taylor & Francis Group
New York   London

First published by The Haworth Press, Inc.

This edition published 2013 by Routledge

Routledge
Taylor & Francis Group
605 Third Avenue, New
York, NY 10017

Routledge
Taylor & Francis Group
2 Park Square, Milton Park
Abingdon, Oxon OX14 4RN

Routledge is an imprint of the Taylor & Francis Group, an informa business

Engineering Libraries: Building Collections and Delivering Services has been co-published simultaneously as *Science & Technology Libraries*™, Volume 19, Numbers 3/4 2001.

Notice:
Product or corporate names may be trademarks or registered trademarks and are used only for identification and explanation without intent to infringe.

Cover design by Thomas J. Mayshock Jr.

Library of Congress Cataloging-in-Publication Data

Engineering libraries : building collections and delivering services / Thomas W. Conkling, Linda R. Musser, editors.
    p. cm.
    Co-published simultaneously as Science & technology libraries, v. 19, nos. 3/4, 2001.
    Includes bibliographical references and index.
    ISBN 0-7890-1672-9 (alk. paper)–ISBN 0-7890-1673-7 (pbk. : alk. paper)
    1. Engineering libraries. 2. Engineering–Computer network resources. 3. Technical librar-ies–Collection development. I. Conkling, Thomas W., 1949- . II. Musser, Linda R. III. Science & tech-nology libraries.

Z675.E6 E54 2001
026.62–dc21
                                                 2001051724

ISBN 13: 978-0-7890-1673-7 (pbk)

Engineering Libraries: Building Collections and Delivering Services

CONTENTS

ABOUT THE EDITORS

Thomas W. Conkling, BS, MLS, is Head of the Engineering Library at The Pennsylvania State University. From 1975-1981 he was Assistant Librarian at the Princeton University Plasma Physics Laboratory Library. He earned a BS in physics and mathematics from the State University of New York at Stony Brook and an MLS from Queens College of the City University of New York. His research interests include the generation, acquisition, and use of technical information, and he has published and given presentations in this area. He is active in the American Society for Engineering Education (ASEE) Engineering Libraries Division. He was co-recipient of the 1995 Engineering Information/Special Libraries Association Engineering Librarian of the Year Award. He also received the 1999 ASEE Engineering Libraries Division Award for Best Paper, and the Homer I. Bernhardt Distinguished Service Award for 2000 from the same group.

Linda R. Musser, BS, MS, is Head of the Earth and Mineral Sciences Library at The Pennsylvania State University. She joined the faculty at Penn State in 1985 as Engineering Reference Librarian/Engineering and Earth Sciences Cataloger prior to attaining her present position in 1991. She earned a BS in civil engineering from the University of Illinois in 1981 and an MS in library and information science from the university in 1985. Her research has focused on the specialized information resources of engineering and the preservation of scientific literature. She is active in the Engineering Libraries Division of the American Society for Engineering Education, the Geoscience Information Society, and the Special Libraries Association. She is a member of Tau Beta Pi.

Introduction

The engineering literature is notable for its sheer size and for the multiple formats it encompasses. Engineers, researchers, and engineering students present a challenging user group because of their diverse and often interdisciplinary information needs. The articles in this volume examine current practices for providing access to engineering information resources, both print and electronic, and look at website development, and instructional and management issues in engineering library settings. Using experience drawn primarily from academic engineering libraries, these papers transcend that venue to provide insights applicable across the sciences and to corporate settings.

The volume is divided into four sections–resources, digital and virtual libraries, information competencies, and management. A key and costly resource for all science and engineering libraries is the scholarly journal. In his paper, Steven Gass reviews the rapidly evolving developments in scholarly communications and makes recommendations for a new communications model. As engineering programs are established and grow it is critical for libraries to follow suit, building collections to support the needs of their clientele. Building an engineering collection from scratch can be a daunting task though, but Beth Brin's paper presents a logical plan for developing book and journal collections to support new engineering programs. Since the volume of publications in engineering make it impossible for a single library to satisfy all user needs with their local collection, access to outside sources is a necessity. Charlotte Erdmann provides a detailed discussion of resource sharing through consortial agreements, interlibrary loan, and document delivery with a focus on the special needs of engineers. One particular area of need for engineers is grey literature. Larry Thompson looks in detail at strategies for acquiring access to technical reports, standards, industrial catalogs, and related types of information. International publications, including technical reports, can be particularly hard to locate and acquire. Bonnie Osif's article looks at the importance of these materials and presents two case studies on international acquisitions strategies by national libraries. The last decade has witnessed an

[Haworth co-indexing entry note]: "Introduction." Conkling, Thomas W., and Linda R. Musser. Co-published simultaneously in *Science & Technology Libraries* (The Haworth Information Press, an imprint of The Haworth Press. Inc.) Vol. 19, No. 3/4. 2001. pp. 1-2; and: *Engineering Libraries: Building Collections and Delivering Services* (ed: Thomas W. Conkling, and Linda R. Musser) The Haworth Information Press, an imprint of The Haworth Press, Inc., 2001. pp. 1-2. Single or multiple copies of this article are available for a fee from The Haworth Document Delivery Service [1-800-342-9678, 9:00 a.m. - 5:00 p.m. (EST). E-mail address: getinfo@haworthpressinc.com].

increased effort at attracting young people to careers in engineering, and resources that can be used to support these efforts are reviewed in the article by Justina Osa and Steven Herb.

The virtual and digital aspects of libraries are becoming increasingly important. Jill Powell provides a detailed description and comparison of the major virtual engineering libraries. As corporate and academic engineering libraries rely more heavily on electronic and Web-based resources, William Mischo provides an insightful review of the progress that is being made towards the true digital engineering library. Since the library website is becoming a crucial interface between libraries and their customers, Christy Hightower's paper presents state-of-the-art technologies and strategies for developing highly functional and useful websites.

Universities around the world graduate tens of thousands of engineers each year but are these engineers graduating with the information seeking skills necessary to succeed in their jobs? Improving user information seeking skills is an ongoing challenge for engineering librarians. Fortunately, the Web presents new opportunities to instruct and educate these students. Leslie Reynolds describes her experiences with developing and teaching a Web-based engineering information literacy course at Purdue University while Ron Rodrigues takes a look at the information skills needed by engineers in the corporate environment, and provides some suggestions for minimum competencies.

The articles in the last section examine several management issues of interest to librarians and administrators. Competition in the global business arena is forcing corporate libraries to rethink their operations and information services. Robert Schwarzwalder's article describes the refocusing of corporate engineering libraries and their use of information technology to better support clientele. The same technology that allows libraries to provide services to users allows Web-based information services to vie for their attention as well. Because of this environment, it is imperative that libraries keep attuned to users' needs. Deborah Helman and Lisa Horowitz present strategies for assessing user needs and relate their experiences with assessment within the MIT library system. Another perpetual need of science and engineering libraries is funding, which seems to be a problem even in good economic times. Exploring external sources for funds is one way to improve the bottom line; and the article by Joanne Lerud and Lisa Dunn provides a guide to all aspects of fundraising for librarians and libraries.

The challenges facing engineering librarians are essentially the same as those of the past–building collections and delivering services that meet the needs of our users. This collection of papers highlights some of the issues, resources, tools and techniques that will be necessary to meet the challenges of engineering librarianship in the future.

Thomas W. Conkling
Linda R. Musser

RESOURCES

Transforming Scientific Communication
for the 21st Century

Steven Gass

SUMMARY. Since its inception in the 17th century the research journal emerged as the formal communication method in the sciences. The last half of the 20th century has seen stresses develop on the journal system due to the explosion of scientific research, increasing subscription costs, and technological advances. New models, taking advantage of digital technology, have demonstrated that great improvements are possible if the scientific community is willing to embrace change. Two methods for significantly changing the model are suggested: adopting an e-print moderator model which decouples the dissemination of information from its review, and shifting the costs of publication from the reader to the author and sponsoring agencies and organizations. *[Article copies available for a fee from The Haworth Document Delivery Service: 1-800-342-9678. E-mail address: <getinfo@haworthpressinc.com> Website: <http://www. HaworthPress.com> © 2001 by The Haworth Press, Inc. All rights reserved.]*

Steven Gass, BS, MLS, is Head, Science and Engineering Libraries, Massachusetts Institute of Technology. He was previously Engineering Librarian and Head of the Science and Engineering Resource Group at Stanford University (E-mail: sgass@ mit.edu).

[Haworth co-indexing entry note]: "Transforming Scientific Communication for the 21st Century." Gass, Steven. Co-published simultaneously in *Science & Technology Libraries* (The Haworth Information Press, an imprint of The Haworth Press, Inc.) Vol. 19, No. 3/4. 2001. pp. 3-18; and: *Engineering Libraries: Building Collections and Delivering Services* (ed: Thomas W. Conkling, and Linda R. Musser) The Haworth Information Press, an imprint of The Haworth Press, Inc., 2001, pp. 3-18. Single or multiple copies of this article are available for a fee from The Haworth Document Delivery Service [1-800-342-9678, 9:00 a.m. - 5:00 p.m. (EST). E-mail address: getinfo@haworthpressinc.com].

3

KEYWORDS. Scientific communication, scholarly communication, scientific journals, e-prints

INTRODUCTION

Much has been written in the last half century about ways to improve the communication of research findings in science and technology. In 1963 a U.S. Presidential Advisory Committee issued a number of recommendations intended to strengthen the communication system for science and technology.[1] Since then the rising cost for information, particularly scientific and technical journals, has exacerbated the problem.[2] The library community has been vocal in recommending solutions to the existing crisis,[3] and it has been joined by many university faculty and administrators in its plea for reform.[4] This paper briefly reviews the factors creating the crisis in scientific communication, examines some of the most promising initiatives that have emerged during the last decade, and recommends two principles any solution must incorporate.

BACKGROUND

Communication lies at the heart of research.[5] It is a key component in the advancement of science.[6] The emergence of the journal as the efficient method for formal communication in the sciences dates back to the 17th century.[7] Despite its success, problems have been associated with journals: delays in publication due to the time involved in the peer review process, constraints on the length of papers, and packaging papers of interest with papers not of interest.[8] Nevertheless the journal system remains the de facto archive for scientific communication, and scientists continue to consider scholarly journals to be extremely valuable.[9] The threat to this archival system results from a combination of forces which now threaten the roles and blur the lines between the various stakeholders in the scholarly communication process: creators (faculty and other researchers), publishers (commercial and societal), and enablers (universities, companies, governments, and libraries).

Three forces have created the current crisis. The first is the rapid growth in scientific knowledge. Since the 17th century science has grown exponentially.[10] Although it now appears that the rate of growth of the formal scientific literature is declining, the total growth of research information being put into circulation annually, formally and informally, remains formidable.[11]

The second force at work has been the commercialization of scholarly publication in the sciences.[12] The years 1950-1975 saw a marked growth in the importance of commercially published journals.[13] This growth contributed to what has long been known in the library community as the serials crisis, which highlighted concerns that a long-term solution requires a fundamental reconfiguration of the dynamics of scholarly communication.[14] A more recent analysis of journal pricing indicates that mergers among commercial publishers have contributed significantly to ongoing price increases.[15]

The third force at work is technology. Information technology is advancing at a rapid pace and becoming ubiquitous.[16] Predictions that technology can solve the crisis in scholarly communication abound.[17] The demise of the traditional scholarly journal has been predicted[18] and celebrated[19] as inevitable due to technological advances. However, ensuring the long-term preservation of digital materials at a level equal to or greater than paper has achieved in the past is far from certain.[20] Although computer centers have demonstrated that archiving digital data is possible,[21] there is still much uneasiness about archiving electronic journals.

In addition to these three forces, it is important to understand how authors' motives and copyright policies contribute to the current situation.

> A common misconception is that people create materials primarily for the fees and royalties that they generate ... Creators whose motive is not financial usually benefit from the widest possible exposure of their work. This creates a tension with their publishers, whose business model is usually to allow access only after payment. Academic journals are an important category of materials in which authors' interests (recording their research and enhancing their reputations, both of which benefit from broad dissemination) may be in direct conflict with the publisher's desire for revenue.[22]

In the current system many scientists do research supported by university resources and government funding, and publish the results of that research. Research articles are submitted either to nonprofit academic publishers or commercial publishers. In return for their services publishers typically demand that authors give them copyright ownership of their articles. Publishers then sell these articles back to the scholarly community, either to libraries or to individuals, as journals. In this cycle of publication universities and their libraries expend large amounts of resources at both the front and back end.[23]

The production and consumption of scholarly information in the academic community has been governed by a gift culture marked by faculty members giving away their research to publishers and expecting to be able to access it for free in their libraries. The commercialization of scholarly publishing over the last fifty years has created a dramatic shift from nonprofit to for-profit publishing, creating a hybrid gift/market system. This has been a major contributor to the current crisis caused by the rising costs of scholarly information.[24]

Scientists value library subscriptions to journals because library subscriptions save them time and money, and help them improve the quality of their work.[25] Library budgets have been unable to keep pace with the increasing volume and cost of scientific scholarship.[26] These increasing costs, combined with the current funding situation for academic institutions and the changing communication infrastructure, require a fundamental redefinition of how libraries operate and cooperate in order to contribute effectively to the scholarly communication process.[27]

So what will happen to the scientific journal? Can the status quo be changed in a manner that takes advantage of new advances without harming the careful archival process that has been developed over the ages? Can the promise of cheaper costs and broader access be realized? Developments during the last decade have begun to reveal some answers.

A DECADE OF PROGRESS

XXX Archive

> In many highly competitive, fast-moving fields of basic science, such as molecular biology, the machinery of publication in standard journals moves too slowly to serve fully the needs of the scientific community. It has therefore become customary for scientists to circulate preprints of articles among their colleagues. Such informal circulation, which harks back to the earliest days of science when new results were communicated by personal letter, has the advantage of speed. But it also has within it the seeds of serious disorders for science . . . The scientific community must devise ways of retaining the timeliness of the preprint and yet reducing its privateness and irresponsibility.[28]

In 1991 Paul Ginsparg at the Los Alamos National Laboratory began an e-print archive for high energy physics which soon expanded to other areas of physics and even to other disciplines.[29] XXX,[30] as the archive is known, remains a vital part of the scientific communication process in physics, and serves as an important model for how the communication process might change in other disciplines.[31] Building on the pre-existing "preprint culture" in high energy theoretical physics which had already become the primary means of communication between researchers in the field, Ginsparg provided a paradigm for improving scientific communication. Using computing and networking advances he provided timely and widely distributed access to important research work. His model has spawned other successful discipline specific e-print archives such as CogPrints[32] for cognitive sciences and RePEc[33] for economics.

Reclaiming Copyright

Responding to perceived unfairness in the copyright system, in 1993 librarians associated with North Carolina's Research Triangle universities led an effort to develop a new publishing and copyright policy:

> As a non-profit institution which relies heavily on government and foundation grants to support its research activities, this university asks its faculty to publish their scientific and technical research results in journals supported by universities, scholarly associations, or other organizations sharing the mission to promote widespread, reasonable-cost access to research information. Where this is not possible, faculty should use the model "Authorization to Publish" form below to ensure that control of copyright in the published results of their university research remains within the academic research community.[34]

This movement toward taking control of one's intellectual output has had some effect. The Association of Computing Machinery copyright policy[35] retains traditional copyright transfer to the ACM but allows authors greater flexibility than in the past, including the right to mount their material on private servers. The American Physical Society has adopted a similar policy.[36]

Highwire Press

In early 1995 Stanford University Library's Highwire Press began online production of the *Journal of Biological Chemistry*, the highly cited journal of the American Society for Biochemistry and Molecular Biology. Soon to follow and partner with Highwire were *Science* and the *Proceedings of the National Academy of Sciences*. Now, in 2000, Highwire produces over 200 journal titles, primarily in the medical and life sciences. Highwire was founded to ensure that its partners, scientific societies and "responsible" publishers, would remain strong and able to lead the transition toward the use of new technologies for scientific communication.[37] It has played a significant role in improving the functionality of electronic journals, pioneering the use of links between authors, articles, and citations, providing advanced searching capabilities, high-resolution images and multimedia, and interactivity. While Highwire should be credited with providing a good model for how to use technology to improve the capabilities of electronic journals, it has had less influence in improving the economic models for the distribution of scientific information. These functional improvements, however, have served to demonstrate how technology can contribute to a positive transformation rather than simply a modernization of the scientific journal.[38]

Scholars' Forum

In March 1997 Caltech held a Conference on Scholarly Communication. At that meeting, attended by university administrators, faculty, and librarians from across the U.S., a consensus emerged that the certification of scholarly articles through peer review could be "decoupled" from the rest of the publishing process, and the peer review process could continue to be supported by the universities whose faculty serve as editors, members of editorial boards, and referees.

> The central idea would have the learned societies expand their role to undertake a certification process for articles, independently of whether they are submitted for, or are eventually published in the standard paper journal system. Under such a system, scholars could submit articles for review (with an agreed-upon submission fee), and the normal refereeing process of the learned society would determine whether the article qualified for their "seal of approval," which, if received, could be affixed to any electronic version of the article as retrieved by others.[39]

The proposal *Scholars' Forum: A New Model For Scholarly Communication*[40] calls for a trilateral partnership between a consortium of universities, professional societies, and authors which would:

1. Support peer review and authentication
2. Support new models of presentation incorporating network technology
3. Permit "threaded" online discourse
4. Adapt to varying criteria among disciplines
5. Assure the security of data
6. Reduce production time and expense
7. Include automated indexing
8. Provide multiple search options

SPARC

The Scholarly Publishing and Academic Resources Coalition (SPARC) was created in 1998. SPARC is a coalition of libraries, initiated by the Association of Research Libraries (ARL), that seeks to partner with scholarly publishers willing to enter markets where journal prices are highest and competition is needed. Through its activities, SPARC intends to reduce the risks to publisher-partners of entering the marketplace and to provide faculty with prestigious and responsive alternatives to current publishing vehicles. Since its inception SPARC has partnered with a number of scholarly societies including the American Chemical Society, Royal Chemical Society, Geological Society of America, and the IEEE.[41] It has also provided grant monies to university initiatives such as Columbia University Press' Earthscape,[42] MIT Press' CogNet,[43] and the California Digital Library's eScholarship.[44] Possibly its most ambitious project to date is BioOne, which will aggregate, link and make easily accessible peer-reviewed research in the biological, ecological and environmental sciences. It hopes to enable leading non-profit journals self-published by scientific societies to remain viable, and offer them a cost-effective alternative to commercial publishers' digital aggregations.[45]

Can SPARC succeed in supplanting journals which are too expensive with less expensive alternatives? While it is having some initial success in introducing significant new journals, in the short term it has created a quandary for libraries which now find themselves having to subscribe to additional titles while convincing faculties that the more expensive titles are no longer necessary. Time will judge whether SPARC will be successful in eliminating, reducing the cost, or making less important the more expensive titles.[46]

NEAR

In October 1998 David Shulenburger, Provost at the University of Kansas, unveiled his proposal for a National Electronic Article Repository (NEAR):

> We must find a way of requiring that when a manuscript prepared by a U.S. faculty member is accepted for publication by a scholarly journal, a portion of the copyright of that manuscript be retained for inclusion in a single, publicly accessible repository, after a lag following publication in the journal . . . NEAR would see to it that articles are permanently archived, thereby assigning responsibility for the solution to another problem brought to us by the electronic age. NEAR could be funded by universities through "page charges" per article included, by federal appropriation, by a small charge levied on each user upon accessing articles or by a combination of these methods.[47]

E-BIOMED

In May 1999 Harold Varmus, then Director of the National Institutes of Health, proposed E-BIOMED: A Proposal for Electronic Publications in the Biomedical Sciences. Within four months this controversial proposal morphed into PubMed Central: An NIH-Operated Site for Electronic Distribution of Life Sciences Research Reports.[48] PubMed Central has created vigorous debate[49] and responses from the private sector.[50] The July 2000 *Freedom of Information Conference: The Impact of Open Access on Biomedical Research*, held at the New York Academy of Medicine, represents the most recent round of debate on this topic at the writing of this article.[51] So far the reaction to

Varmus' proposal demonstrates the problems associated with the existing copyright structure,[52] a copyright structure which has been described as a Faustian bargain made by scholars to get their work published.[53]

Open Archives Initiative

In October 1999 digital librarians and computer scientists committed to creating a Universal Preprint Service (now known as the Open Archives Initiative[54]) gathered in Santa Fe, New Mexico, for a meeting sponsored by the Council on Library and Information Resources, the Digital Library Federation, SPARC, ARL, and the Research Library at the Los Alamos National Laboratory. The objective of the meeting was to pave the way for universal public archiving of the scientific and scholarly research literature on the Web.[55] Participants concluded that many different archive initiatives were likely to emerge, and, for such initiatives to become part of the scholarly communication system, interoperability was essential. Further, a consensus developed that interoperability hinges on a clear distinction between the archive functions and end-user functions.[56]

Although many technical issues were identified and discussed, social issues concerning scholarly communication also emerged:

1. Will the institution provide or support a departmental or institutional e-print archive of authors associated with the institution? If so, will it adopt the open archive principles agreed to in Santa Fe?
2. How will research libraries package and deliver access to e-print literature?
3. With the resolution of e-print archive interoperability technical issues, what will be the process of resolving the social issues connected with tenure and publishing?[57]

Building Consensus

In March 2000 a conference was held in Tempe, Arizona, sponsored by the Association of American Universities, the Association of Research Libraries, and the Merrill Advanced Studies Center of the University of Kansas.[58] Participants produced a set of *Principles for Emerging Systems of Scholarly Publishing*.[59] In the preface to their principles they state:

> Numerous studies, conferences, and roundtable discussions over the past decade have analyzed the underlying causes and recommended solutions to the scholarly publishing crisis. Many new publishing models have emerged. A lack of consensus and concerted action by the academic

community, however, continues to allow the escalation of prices and volume.

Their hope is to build consensus on a set of principles that will inform the design and evaluation of new systems of scholarly publishing. The goal is to provide guidance while leaving open to creativity and market forces the actual development of such systems.

DSpace

In July 2000 Hewlett Packard and the MIT Libraries began a two year research project called DSpace.[60] DSpace aims to develop a scalable digital archive with storage, submission, retrieval, searching, access control, rights management, and publishing capabilities. Its goal is to embrace all of the digital intellectual output of MIT. DSpace is one of the first e-print archive initiatives to focus on capturing an institution's output rather than a particular field of scholarship, and, in addition, it is one of the first involving an industry partner, HP.[61] It hopes to contribute answers to some of the questions raised by the Open Archives Initiative.

eScholarship

In July 2000 the University of California launched eScholarship,[62] dedicated to facilitating scholar-led innovations in scholarly communication. The overall goal of eScholarship activities is to develop an infrastructure for digitally based scholarly communication that:

1. Facilitates the mutual interests of the University, its faculty, and the broader scholarly community.
2. Leverages the capabilities and strengths of UC to provide leadership in this area.
3. Supports and extends experimental reconfigurations of the components of scholarly communication by communities of scholars themselves.

eScholarship sees itself as a core set of disciplinary e-print archives surrounded by functional tools and orbited by value-added scholarly "products" created by scholarly societies, the university, or third parties. It includes the following key components:

1. Disciplinary-based knowledge archives of e-prints.
2. Support tools for submission, expanded peer review, discovery, access, and use.
3. New scholarly products drawn from e-print archives.[63]

E-Prints for Chemistry?

At the August 2000 American Chemical Society National Meeting in Washington, D.C., the ACS sponsored a Webcast *Online Preprints: Implications for Chemistry*.[64] Although no consensus was reached, the points raised at this forum effectively highlight many of the questions and issues surrounding the debate over e-prints:

1. The need for effective, efficient means of information dissemination that uses all available technology to reach as many readers as possible.
2. The fact that disciplines have different cultures, e.g., physics and math researchers use preprints as an integral form of communication.
3. The need to determine whether preprint articles are considered "prior publications" and what that means for current society practices for journal publication.
4. The issue of whether proper content and credit fairly applied can be assured without peer review.
5. The relationship between preprints and patent applications.
6. Organizational questions: How to name, archive, and file preprint information to allow for future access.

CHANGING THE PARADIGM

XXX represents the most significant change in scientific communication since the establishment of the journal in the 17th century. Although not intended to replace journals, it quickly became used as an electronic journal for obvious reasons of convenience.[65] XXX has shown that simply focusing on the status quo of scientific journal creation and dissemination is shortsighted. It demonstrates that the basic distribution of scholarly papers can be achieved inexpensively while at the same time increasing access.[66] Can its success be transferred to other scientific communities? Or are the cultural differences between disciplines too great a barrier? The World Wide Web is an example of how an innovation developed by the high energy physics community has been quickly adapted by the world at large.[67] Why not e-prints?

The momentum created by XXX, the Open Archives Initiative, and other related projects is spawning a transformation in scientific communication. Clearly the dissemination of scholarly work can be separated from the peer review process. Although the peer review process is not perfect,[68] most scholars, including physicists, value its importance.[69] The American Physical Society has successfully begun to work with XXX to speed up peer review.[70] In addition the initial concern some authors displayed regarding the lack of prestige associated with scholarly electronic journals is eroding.[71]

Serious obstacles remain, however, and one of the largest is convincing publishing stakeholders to change their financial models. Commercial publishers and many scholarly societies have benefited financially from the current model for scientific journals. They are grappling with their mission of advancing knowledge in their respective disciplines versus the dependence they have developed on the income generated by their publishing programs. Another challenge is the social and cultural differences between the different scientific communities, as demonstrated by the previously discussed E-BIOMED proposal and the ACS Webcast on Online. Social issues are often more difficult than technical ones, and can be hard to overcome.[72] Although scientists pride themselves on their belief in logic and rational thought, old traditions do die hard and university communities are not noted for rapid change.

I believe the most promising strategy for improving the flow of scientific communication involves a combination of two ideas. The first is moving to an e-print moderator model [see Figure 1]. This has the potential to allow the widest range of scientific manuscripts to be archived, searched, and distributed electronically with the lowest possible cost.[73] It will take cooperation between authors, editors, reviewers, societies, universities, and responsible publishers.

Second, the costs for publication must shift from the reading community to authors and the funding agencies/institutions sponsoring their research either by direct support or through overhead costs.[74] In fact it is argued that open access publishing is becoming a permanent feature of the Internet, and that an economic model is emerging where the costs are paid for by the suppliers of the information.[75] Certainly the costs borne by current subscription models are counterproductive to the goal of disseminating scholarly work as far as possible. For years librarians and others have decried the increasing costs of journals and been forced to cancel subscriptions, thus impeding access to scholarly work. By reverting to a funding system more akin to page charges, authors and the organizations sponsoring their research will be more aware of the costs associated with disseminating their work, and will be more responsible in determining where their work is published. In their exhaustive work on electronic journals Tenopir and King comment that reverting to a page charge system would be a great gain from a system economic standpoint.[76] Although they acknowledge that the basic research funders would bear a larger burden of the costs associated with publishing, they go on to state "society would benefit because more use, and, therefore, value would be derived from the basic research findings." Is this not a goal worth fighting for?

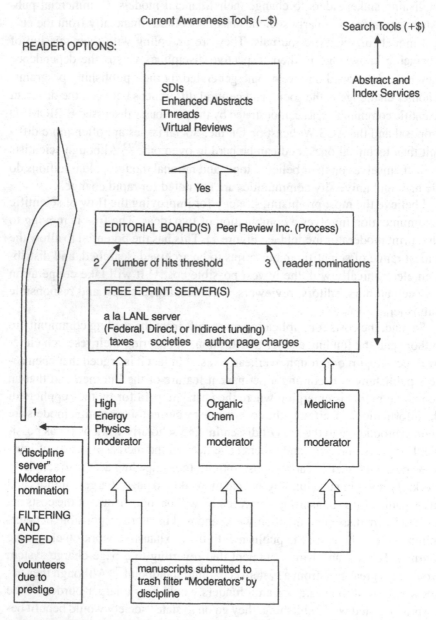

FIGURE 1. E-Print Moderator Model

From *Guide to Information Sources in the Physical Sciences* by David Stern ©2000 Libraries Unlimited. (800) 237-6124 or www.lv.com. Reprinted with permission.

REFERENCES

1. United States President's Science & Advisory Committee (PSAC), *Science, Government, and Information: The Responsibilities of the Technical Community and the Government in the Transfer of Information* (Government Printing Office, Washington, D.C., 10 January 1963).

2. Anthony M. Cummings et al., *University Libraries and Scholarly Communication* (The Association of Research Libraries for the Andrew W. Mellon Foundation, November 1992), 88.

3. Joseph J. Branin and Mary Case, "Reforming Scholarly Publishing in the Sciences: A Librarian Perspective," *Notices of the AMS*, 45: 4 (April 1998), 475-486.

4. Denise K. Magner, "Seeking a Radical Change in the Role of Publishing," *Chronicle of Higher Education* 46:41 (June 16, 2000), A16.

5. A. J. Meadows, "Communicating Research," Academic Press, San Diego, 1998, p. ix.

6. I. Bernard Cohen, *Revolution in Science* (Belknap Press of Harvard University Press, Cambridge, MA, 1985), 27-33.

7. Meadows, 7.

8. Bernard Houghton, *Scientific Periodicals: Their Historical Development, Characteristics and Control* (Linnet Books & Clive Bingley, Great Britain, 1975), 42-51.

9. Carol Tenopir and Donald W. King, *Towards Electronic Journals: Realities for Scientists, Librarians, and Publishers* (Special Libraries Association, Washington, D.C., 2000), 26.

10. Derek J. De Solla Price, "Prologue to a Science of Science," *Little Science, Big Science* (Columbia University Press, NY, 1963), 1-21.

11. Meadows, 31-32.

12. Branin, 478-481.

13. Brian Vickery, "A Century of Scientific and Technical Information," *Journal of Documentation* 55:5 (December 1999), 497.

14. Cummings, 94-99.

15. Mark J. McCabe, "Academic Journal Pricing and Market Power: A Portfolio Approach," *Conference on the Economics and Usage of Digital Library Collections*, (Ann Arbor, MI, 23-24 March 2000), http://www.si.umich.edu/PEAK-2000/mccabe.pdf.

16. Christine L. Borgman, *From Gutenberg to the Global Information Infrastructure* (MIT Press, Cambridge, MA, 2000), 1.

17. Borgman, 81-92.

18. Andrew M. Odlyzko, "Tragic loss or good riddance? The impending demise of traditional scholarly journals," *International Journal of Human-Computer Studies* 42 (1995), 71-122, http://www.research.att.com/~amo/doc/tragic.loss.long.pdf.

19. Stevan Harnard, "Free at Last: The Future of Peer-Reviewed Journals," *D-Lib Magazine* 5:12 (December 1999), http://www.dlib.org/dlib/december99/12harnad.html.

20. Borgman, 202-3.

21. Clifford Lynch, *MIT Communications Forum on Digital Libraries*, (April 20, 2000), http://media-in-transition.mit.edu/forums/library/index_summary.html.

22. William Y. Arms, *Digital Libraries* (MIT Press, Cambridge, MA, 2000), 82-3.

23. Branin, 481.

24. Peter Lyman, "Digital Documents and the Future of the Academic Community," *Technology and Scholarly Communication* (University of California Press published in association with The Andrew W. Mellon Foundation, Berkeley, 1999), 366-379 (368-369).

25. Tenopir, 218-219.

26. Branin, 477.

27. Brian L. Hawkins, "The Unsustainability of the Traditional Library and the Threat to Higher Education," in *The Mirage of Continuity: Reconfiguring Academic Information Resources for the 21st Century* (Council on Library and Information Resources and Association of American Universities, Washington, D.C., 1998), 129-153.

28. United States President's Science & Advisory Committee, 20.

29. Paul Ginsparg, "First Steps Towards Electronic Research Communication," *Computers in Physics* 8:4 (July/August 1994), 390-6.

30. arXiv.org e-Print archive [cited December 5, 2000]. http://arXiv.org/.

31. Gary Taubes, "Electronic Preprints Point the Way to 'Author Empowerment,'" *Science* 271:9 (February 1996), 767-8.

32. Cogprints: Cognitive Sciences Eprint Archive [cited December 5, 2000]. http://cogprints.soton.ac.uk/.

33. RePEc: Research Papers in Economics [cited December 5, 2000]. http://repec.org/.

34. TRLN Model University Policy Regarding Faculty Publication in Scientific and Technical Scholarly Journals [cited December 5, 2000]. http://www.lib.ncsu.edu/scc/trln.html#policy.

35. ACM Interim Copyright Policy, Version 3, 12/18/98 [cited December 5, 2000]. http://www.acm.org/pubs/copyright_policy/.

36. American Physical Society Transfer of Copyright Agreement, revised 6/00 [cited December 5, 2000]. ftp://aps.org/pub/jrnls/copy_trnsfr.asc.

37. Highwire Press: A Brief Introduction [cited December 5, 2000]. http://highwire.stanford.edu/intro.dtl.

38. Clifford Lynch, "The Transformation of Scholarly Communication and the Role of the Library in the Age of Networked Information," *The Serials Librarian* 23:3-4 (1993), 5-20 (7-9).

39. Charles E. Phelps, *The Future of Scholarly Communication: A Proposal for Change* (June 5, 1997) [cited December 5, 2000]. http://www.econ.rochester.edu/Faculty/PhelpsPapers/Phelps_paper.html.

40. Anne M. Buck, Richard C. Flagan, and Betsy Coles, *Scholars' Forum: A New Model For Scholarly Communication* (California Institute of Technology, Pasadena, CA, March 23, 1999), [cited December 5, 2000]. http://library.caltech.edu/publications/ScholarsForum/.

41. SPARC: The Scholarly Publishing and Academic Resources Coalition [cited December 5, 2000]. http://www.arl.org/sparc/index.html.

42. Columbia Earthscape: An Online Resource on the Global Environment [cited December 5, 2000]. http://www.earthscape.org/.

43. MIT Cognet: The Brain and Cognitive Sciences Community Online [cited December 5, 2000]. http://cognet.mit.edu/.

44. eScholarship: Scholar-led innovations in scholarly community [cited December 5, 2000]. http://escholarship.cdlib.org/.

45. Richard K. Johnson, "A Question of Access: SPARC, BioOne, and Society-Driven Electronic Publishing," *D-Lib Magazine* 6:5 (May 2000) [cited December 5, 2000]. http://www.dlib.org/dlib/may00/johnson/05johnson.html.

46. Mark Doyle, "Surviving the Transition," *Learned Publishing* 13:3 (July 2000), 138, http://www.catchword.com/alpsp/09531513/v13n3/contp1-1.htm.

47. David E. Shulenburger, "Moving With Dispatch to Resolve the Scholarly Communication Crisis: From Here to NEAR," *Association of Research Libraries Membership Meeting Proceedings* (October 1998), http://www.arl.org/arl/proceedings/133/shulenburger.html.

48. PubMed Central: An NIH-Operated Site for Electronic Distribution of Life Sciences Research Reports [cited December 5, 2000]. http://www.nih.gov/about/director/ebiomed/ebi.htm.

49. Barry P. Markowitz, "Biomedicine's Electronic Publishing Paradigm Shift: Copyright Policy and PubMed Central," *Journal of the American Medical Informatics Association* 7:3 (May/June 2000), 222-229 (227).

50. BioMed Central [cited December 5, 2000]. http://www.biomedcentral.com/default.asp.

51. Freedom of Information Conference: The Impact of Open Access on Biomedical Research (New York Academy of Medicine, July 6-7, 2000) [cited December 5, 2000]. http://www.biomedcentral.com/info/information.asp.

52. Michael W. Jacobson, "Biomedical Publishing and the Internet: Evolution or Revolution," *Journal of the American Medical Informatics Association* 7:3 (May/June 2000), 230-233 (230).

53. Stevan Harnard, "The PostGutenberg Galaxy: How to Get There From Here," *Information Society* 11(4), 285-292, http://cogsci.soton.ac.uk/~harnad/THES/thes.html.

54. Open Archives Initiative [cited December 5, 2000]. http://www.openarchives.org.

55. Richard E. Luce, "The Open Archives Initiative," *College & Research Libraries News* 61:3 (March 2000), 184-186, 202 (185).

56. Luce, 186.

57. Luce, 202.

58. Magner, A16.

59. Principles for Emerging Systems of Scholarly Publishing (May 10, 2000) [cited December 5, 2000]. http://www.arl.org/scomm/tempe.html.

60. DSpace [cited December 5, 2000]. http://libraries.mit.edu/dspace.

61. DSpace: Frequently Asked Questions [cited December 5, 2000]. http://web.mit.edu/dspace/about_us/faqs/index.html.

62. eScholarship: Scholar-led innovations in scholarly community [cited December 5, 2000]. http://escholarship.cdlib.org/.

63. About eScholarship [cited December 5, 2000]. http://escholarship.cdlib.org/about.html.

64. Online Preprints: Implications for Chemistry (August 20, 2000) [cited December 5, 2000]. http://archive.netpodium.com/acs/082000.

65. Ginsparg, 393.

66. Mark Doyle, 138-9.

67. Borgman, 98.

68. Meadows, 183-194.

69. James Langer, "Physicists in the New Era of Electronic Publishing," *Physics Today* 53:1.1 (August 2000), 36-7, http://www.aip.org/pt/vol-53/iss-8/p35.html.

70. Arthur P. Smith, "The Journal as an Overlay on Preprint Databases," *Learned Publishing* 13:1 (January 2000), 43-48 (44), http://www.catchword.com/alpsp/09531513/v13n1/contp1-1.htm.

71. Michael Fosmire and Song Yu, "Free Scholarly Journals: How Good Are They," *Issues in Science and Technology Librarianship* 27 (Summer 2000) [cited December 5, 2000]. http://www.library.ucsb.edu/istl/00-summer/refereed.html.

72. John Seely Brown and Paul Duguid, *The Social Life of Information* (Harvard Business School Press, Boston, 2000), 40.

73. David E. Stern, *eprint Moderator Model* [cited December 5, 2000]. http://www.library.yale.edu/scilib/modmodexplain.html.

74. Thomas J. Walker, "The Future of Scientific Journals: Free Access or Pay Per View," *American Entomologist* 44 (Fall 1998), 135-138, http://csssrvr.entnem.ufl.edu/~walker/fewww/aecom3.html.

75. Tenopir, 401.

76. William Y. Arms, "Economic Models for Open Access Publishing," *iMP: Information Impacts Magazine* (March 2000) [cited December 5, 2000]. http://www.cisp.org/imp/march_2000/03_00arms.htm.

Building a Library Collection to Support New Engineering Programs

Beth L. Brin

SUMMARY. This article describes strategies for developing a collection to support engineering programs that are new to the campus. The focus is on small to medium-sized institutions and the primary materials covered are reference sources; current and retrospective monographs; serials including periodicals, conference proceedings and annuals; and other specialized materials. Discussions of the environment, funding, assessment, and the possible impact on other library services are also included. *[Article copies available for a fee from The Haworth Document Delivery Service: 1-800-342-9678. E-mail address: <getinfo@haworthpressinc.com> Website: <http://www.HaworthPress.com> © 2001 by The Haworth Press, Inc. All rights reserved.]*

KEYWORDS. Engineering, collection development, new academic programs, college or university libraries

New programs are a natural product of a growing college or university. Often, the added program builds on the strengths of the current offerings of a uni-

Beth L. Brin is Reference and Bibliographic Instruction Librarian, and Collection Development Liaison to the Civil, Electrical, and Mechanical Engineering Departments, and the Construction Management Department at Boise State University, Albertsons Library, Boise, ID 83725-1430 (E-mail: bbrin@boisestate.edu).

The author thanks Tim Brown, Peggy Cooper, Dan Lester, and Gloria Ostrander-Dykstra for their review of the original manuscript.

[Haworth co-indexing entry note]: "Building a Library Collection to Support New Engineering Programs." Brin, Beth L. Co-published simultaneously in *Science & Technology Libraries* (The Haworth Information Press, an imprint of The Haworth Press, Inc.) Vol. 19, No. 3/4, 2001, pp. 19-37; and: *Engineering Libraries: Building Collections and Delivering Services* (ed: Thomas W. Conkling, and Linda R. Musser) The Haworth Information Press, an imprint of The Haworth Press, Inc., 2001, pp. 19-37. Single or multiple copies of this article are available for a fee from The Haworth Document Delivery Service [1-800-342-9678. 9:00 a.m. - 5:00 p.m. (EST). E-mail address: getinfo@haworthpressinc.com].

versity. Occasionally, the new program will open an entirely new direction of academic study for the institution. In each case, the ramifications for the library include assessing the existing collections, purchasing materials to support the new program(s), reviewing current services to ensure that the needs specific to the new discipline are addressed, and developing expertise within the library to accommodate the service and collection needs. It should be understood that programs taking the institution in a new direction will require more resources than those that build on existing strengths. Most new programs do necessitate some additional resources.

The literature provides some discussion of the impact or support of new programs. In the academic environment, Lanier and Carpenter (1994) describe the impact of a decision to regionalize programs in pharmacy and nursing; Yutani (1993) writes of the process involved with developing new Japanese collections and services; Linklater (1988) discusses support for new courses in computer information systems; and Johnson (1999) presents the approach taken when a new environmental and human health program was initiated. Stevens (1991) wrote a related discussion piece on strategies for obtaining additional support for the information needs of new faculty. Articles on building a collection for a new library included those written by Pearson (1999) for an undergraduate library, Myers (1991) for an upper undergraduate campus, Lowman and Blatz (1988) for community college centers and Allen (1998) for a school library. Issues addressing building a new serials collection are discussed by Woodberry (1993).

This article will address building a collection to support new engineering programs for which there was minimal support prior to the program approval. The focus of the article will be small to medium-sized programs and library funding levels. Service issues will also be briefly covered. Although the discussion will address general application, examples will be used from the case of Boise State University Albertsons Library, hereafter referred to as the Library.

In November 1995, the Idaho State Board of Education approved Boise State University's request to confer undergraduate degrees in three engineering disciplines: civil, electrical and mechanical. At that time, Boise State University had a student FTE of 10,338 who had the choice of 63 Baccalaureate, 22 Master's and 1 Doctorate degrees with 875 faculty members (including adjunct professors). The Library held 440,000 volumes, 4,700 serials and had 57 staff members (FTE). The first classes in these new programs were taught in the Fall of 1996. Prior to this decision, coursework for the first two years of an engineering degree were provided. Students completing the coursework would then transfer to an institution granting engineering degrees. Other related programs that the Library had supported are a Construction Management program, Computer Information Systems/Production Management program, and programs in Physics, Chemistry, Geosciences, Computer Science and Applied Technology. In January 2000, the State Board approved Boise State University's request to confer Master's degrees in four areas: civil, computer, electrical, and mechanical engineering. In Fall 1999, Boise State University had a student FTE of 11,330 who had the choice of 90 Baccalaureate, 35 Master's and 2 Doctorate degrees with 935 faculty members (including adjunct professors). The Library held 521,000 volumes, 5,200 serial subscriptions, 2,500 additional serials through full-text databases, and had 61.5 staff members (FTE). Members of the Library staff at Boise State University, including the author, have been building the library collection to support the engineering programs since the first approval was granted.

ENVIRONMENT

The author of this article is making the assumption that the campus library is willing and/or is mandated to support all of the university's programs, including new offerings. The degree of support will depend on the resources provided by the university administration, the library administration's allocation of those funds, and any donated funds or endowments. The resources for the university will be dependent on the funding sources of the institution—for public universities this typically includes a mix of state-appropriated funds, student tuition and/or fees, endowments, grants, and donations. In theory, the campus library should receive additional funding for any new program requiring additional library resources. Each institution will have its own history on this issue, and the library administrators will be very aware of their own cases. Related issues include questions such as: how are new programs approved within the university or college? is the library's input required or requested? if input is not requested, is the library informed about the new programs? how are the needed resources determined and requested? This article will provide some information for those institutions that are preparing justifications for additional resources to support new engineering programs.

The campus environment is important, but the environment within the state may also impact decisions. Is there a state board that approves all new programs? Are other institutions of higher education in the state providing similar programs? Will the state board expect cooperation or collaboration between the state institutions and if so, does that impact the library? Are some of the necessary information resources currently provided through a state or regional consortium? The mix of these variables will be unique for each institution and may provide parameters for later decisions concerning library resources and services. The questions will be slightly different for private institutions, but the regional environment will still impact decisions.

A key advantage of a new program is the opportunity to collaborate with the faculty from the very beginning on the library resources and services needed for that new program. Even if resources are scarce, working with faculty at the inception of a program has the potential of raising awareness about the library's ability to meet information needs in diverse ways. Many faculty members are unaware of the costs of services, inflation rates, licensing agreements and restrictions and this is an opportune time to begin to educate the faculty on these issues. It is important to identify the program's planned offerings by obtaining copies of proposed degree requirements and course descriptions. These program offerings are essential for establishing collection goals. Continuing dialogue with faculty members and others is essential to keep informed about changes that are certain to take place. The administrative structure may change and the focus of the program will be somewhat dependent on the research interests of the new faculty, especially for graduate studies. The dialogue will also help to educate faculty on progress towards library goals. The strategic plan of the new program is also crucial: will the program grow to serve additional disciplines, degrees, or levels of students (undergraduate, masters, doctoral)? what is the projected enrollment? These factors will influence the collection being built and the services being provided.

FUNDING FOR THE LIBRARY

The appropriate level of funding for any program is influenced by the local conditions. If the new program is highly visible within the university and within the region, there may be significant incoming resources, especially if local industry monetarily supports the new programs. Whether any of these resources are applied to library resources and services is another local issue. Startup programs for engineering typically focus on hiring new faculty and constructing appropriate lab facilities. Still, if the program has high visibility, the university administration may be more easily persuaded to allocate funds to support the purchase of library materials and extend appropriate services. One could impress upon the university administration the importance of both one-time funds and continuing funds. One-time funds can help purchase books and serials retrospectively; continuing funds are essential for current books and serials, as well as supporting current and new services.

The amount to include in a funding proposal will vary. Herring's 1992 article discussing the underfunding of engineering programs at southeastern academic institutions presents summary information for engineering and other collection areas that may be useful. It may also be instructive to obtain figures from similar institutions concerning the support they provide for engineering programs. Begin by identifying: (1) institutions that are used for institutional benchmarking studies, (2) institutions that closely match the planned suite of engineering programs, and (3) institutions in the region that offer engineering programs. Contact the library collection development officers of the institutions, and explain the project. They may provide the information themselves, or recommend the appropriate person within the organization for that information. Ask for the library expenditures for books, serials, indexes and other reference resources, and local funding arrangements for packages, consortia purchases, or approval plans. The way library statistics are kept will vary between institutions. Are monographic series considered books or serials? Are the funds budgeted by program/ department, in a lump sum, or some other method? Do serials cover print and electronic? Are indexes funded from a reference fund or from the discipline the index supports? Getting figures that include electronic indexes and electronic journals that are discipline specific is especially problematic since many institutions are buying packages or joining consortia and funds are listed in a lump sum. So the figures received will not be exactly comparable, but they will provide information concerning the range of expenditures by targeted institutions. Also obtain information on the level of the program (undergraduate only, undergraduate and masters, or degrees through doctoral studies), the number of engineering disciplines covered, and the number of students at the various levels. A good resource for these figures is *Profiles of Engineering and Engineering Technology Colleges* published annually by the American Society for Engineering Education (ASEE). Data in this publication is supplied by the institution to ASEE and is not available for every institution in the catalog.

For administrators who value benchmarking efforts, these figures may add validity to the justification request. The faculty may also find the numbers of interest. At Boise State University, the author worked with a committee from the College of Engineering to develop a library funding proposal as part of their charge in detailing the needed support for adding Master's degrees to the engineering program. Some members of the committee were very surprised at the expense involved in purchasing the engineering literature, and the benchmark results convinced them that figures in excess of $100,000 were common.

Another aid in obtaining funds to support the new engineering programs is the importance of accreditation. Although the Accreditation Board for Engineering and Technology (ABET), the organization that accredits engineering and engineering technology undergraduate programs, is changing the process used to review programs, past reviews have had a section that included the library. There are no set guidelines for specific titles or minimum expenditures, and questions asked during the accreditation visit vary depending on the team members, but the accreditation team typically wants to see evidence that the engineering students, staff and faculty are being supported by the library, and that the faculty do have input into materials being provided. Hopefully, this expectation will carry into the new review process that focuses on outcomes.

COLLECTION ASSESSMENT

A primary step in building a collection is assessing the current status of the collection. Some of the same tools used to build a collection can be used to first assess it. For libraries that have not supported engineering in the past, the assessment will show very few holdings. Holdings in selected bibliographies and key indexes can be calculated. (See the following section for specific titles.) Unfortunately, there is no core list of mandated or recommended titles for engineering, either in books or in serials. Noting what types of materials are in the circulating collection, and the age of those materials, is a useful step. Another tool for assessment of the collection is the conspectus approach, applying collection levels to call number areas based on the conspectus classification. This approach identifies strengths and weaknesses within a collection and helps to identify those areas on which to focus collecting activity.

As described, at Albertsons Library there were some related program offerings but no previous engineering degrees conferred. The Library collection included some basic holdings in reference for appropriate indexes, guides to literature, encyclopedias, dictionaries, and handbooks. Much of the circulating collection was fairly old, the result of donations to the collection. The Library had reviewed the collection in 1994 using the WLN Conspectus method and identified most T sections as falling in the 1a or 1b area (minimal coverage). Exceptions were noted in small areas of call number sections TA and TH (supporting Construction Management and Geosciences), TK (supporting Computer Information Systems and some applied technology courses), TN (supporting Geosciences), TS (supporting Production Management), and TX (supporting some applied technology courses). There were a few anomalies such as a high number of titles on space travel and petroleum engineering. Even then, only the TS155-193 (Production Management) area reached a 3a level (basic study–the target for an undergraduate program). The supporting sections in QA, QC and QD fared better since there has been program support for Computer Science, Mathematics, Physics, and Chemistry but as might be expected specific sections such as QC310.15-318 (Thermodynamics) were inadequate. With this assessment, the staff knew it would take a concerted effort over time to build a collection supporting the new programs.

BUILDING THE COLLECTION

Once the planned program offerings are known and the current state of the collection assessed, prepare preliminary collection development guidelines for each subject area, based on the knowledge of the intended program offerings. The guidelines should be incorporated in the library's Collection Development Policy and will be the blueprint for developing the collection. The guidelines can also be used as a tool for discussion with faculty regarding the development of the collection. Remember that the collection development policy is a work-in-progress; revise the guidelines after faculty members have been hired and the program takes on a fuller shape.

Before suggesting some specific strategies, some observations about the engineering literature may be helpful. Engineering itself is an interdisciplinary pursuit, requiring information from the basic sciences, applied science, and/or technologies depending on the project or research at hand. At the same time, engineering is divided into several disciplines, each of which has its own literature. Therefore, some resources that are interdisciplinary will be helpful, and other resources that focus on the discipline or sub-discipline will be necessary. A good discussion of the interdisciplinary nature of engineering is found in Posey's (1987, 79-95) discussion of selecting for engineering disciplines. Each discipline has a major society, and sub-disciplines may also have their own society. Most societies do publish some material–the major societies publish monographs, journals, conference proceedings, standards and other materials. Society publications form an essential part of any definition of the engineering "core literature." Small to medium-sized libraries will probably not be able to purchase all society publications, and some titles are more suitable for graduate collections or practitioners. But be sure to include some society publications in the orders for the collection. Commercial publishers also publish engineering literature–monographs and serials are well-represented and particularly books and textbooks for the undergraduate. Any determination of "core" will vary depending on the programs the institution offers, the focus of the faculty, and the level of degrees offered.

Now let's focus on identifying titles to order for the collection. As previously mentioned, there is no core list of monographs and serials that are needed for the support of engineering programs. Bibliographies related to reference titles and indexes are available but those for the circulating collection are scarce and dated. The following three sections identify bibliographies and collecting strategies for: (1) reference materials, (2) monographs, and (3) serials. A fourth section discusses other materials that are useful for engineering research.

Reference Materials

Reference materials for engineering are covered by a few standard guides to the literature. Most of these include annotated bibliographies for various categories such as abstracts and indexes, encyclopedias, or handbooks. The most recent include:

Hurt, Charlie Deuel. *Information Sources in Science and Technology*. 3rd ed. Englewood, CO: Libraries Unlimited, Inc., 1998.

> This work arranges material by discipline, and then by type of material within the discipline. There are nine chapters covering engineering disciplines. Categories include history, guides to the literature, bibliography, abstracts and indexes, encyclopedias, handbooks, directories, and websites.

Lord, Charles. *Guide to Information Sources in Engineering*. Englewood, CO: Libraries Unlimited, Inc., 2000.

> This source arranges material by type and within specific categories of materials (scholarly journals, handbooks, internet resources, and professional and trade associations). The material is further divided by discipline; up to fifteen disciplines are included. Other categories include general reference sources, information access tools, grey literature, buyer's guides, regulations, government resources, and education and career resources. A chapter discussing how engineers use information introduces the work. Materials listed are English language and primarily published from 1996-1999.

Malinowsky, Harold Robert. *Reference Sources in Science, Engineering, Medicine and Agriculture*. Phoenix: Oryx, 1994.

> The material in this title is arranged by discipline, and then by type of material within the discipline. There are nine chapters covering engineering disciplines, although the breakdown is different than Hurt's arrangement. In addition to the categories included by Hurt, Malinowsky includes standards, tables, treatises, and periodicals.

Mildren, Ken W. and P. J. Hicks. *Information Sources in Engineering*. 3rd ed. London: Bowker-Saur, 1996.

> The approach taken by Mildren and Hicks is quite different than that of Hurt, of Lord, or of Malinowsky. The first section is a fairly lengthy description of the types of sources, and then bibliographic essays of sub disciplines and some disciplines are provided. These bibliographic essays provide a summation of the field's focus and structure, with a description of the literature embedded. Examples of sub disciplines include stress analysis and thermodynamics; examples of disciplines include chemical engineering and nuclear en-

gineering. This source does include reference works of interest as well as organizations, journals, important conferences, and textbooks that are paramount in the field. This guide has a British focus.

Two titles that are now dated but still contain good information about the types of literature in the engineering disciplines are:

Chen, Ching-chih. *Scientific and Technical Information Sources.* 2nd ed. Cambridge, MA: MIT Press, 1987.

Mount, Ellis. *Guide to Basic Information Sources in Engineering.* New York: Wiley, 1976.

For information on the literature of specific disciplines, consult one of the Engineering Literature Guides series, produced by the Engineering Libraries Division (ELD) of the American Society for Engineering Education. Each guide focuses on a discipline, such as Civil Engineering, or a sub-discipline such as Applied Optics. It's worth looking at the list of current guides to determine if any are appropriate for the collection. The current guides are listed at: *http://www.englib.cornell.edu/eld/publications.html*.

These guides to the literature will help in the selection of encyclopedias, dictionaries, indexes, directories, handbooks and other reference material that are critical to supporting the engineering programs.

Let's consider some specific titles from these sources. The collection may already include a standard encyclopedia such as the *McGraw-Hill Encyclopedia of Science and Technology*. Depending on the program offerings, consider selecting specialized encyclopedias such as the *Wiley Encyclopedia of Electrical and Electronics Engineering* (also available as an online subscription), or the *Kirk-Othmer Encyclopedia of Chemical Technology*. Both are broadly applicable to a variety of interdisciplinary topics that relate respectively to electrical engineering or chemical engineering. Ordering a selection of handbooks representing the disciplines covered by the new programs is also recommended. Consider selecting handbooks to represent disciplines for which no program is offered but that may cover topics of tangential interest. For example, Boise State University does not have a chemical engineering program but Perry's *Chemical Engineers' Handbook* is part of the Boise State University Albertsons Library collection. O'Gorman recently published a list of core engineering reference titles in Booklist (1999) that lists key encyclopedias and handbooks.

For indexes, there are a wide variety of specialized indexes available. Powell has written articles about several engineering indexes on CD-ROM format (1991) and an informal study of the CD-ROMs to which libraries supporting engineering subscribed (1993). Although the CD-ROM format being described is outdated, the articles provide useful information about the indexes themselves. Three indexes that cover a broad range of engineering disciplines are H. W. Wilson's *Applied Science and Technology Index* (abstracts available in the online version), Engineering Information's (Ei) *CompendexPlus* (title may vary), and *Science Citation Index* (the web version is *Web of Science*). An undergraduate program may be able to minimally manage with only a subscription to *Applied Science and Technology Index*. However, reliance on electronic access to *CompendexPlus* on a per-use fee for those students who are doing specialized projects, such as their senior design projects, is suggested. Of the three interdisciplinary indexes, only *CompendexPlus* regularly indexes conference papers, an important category of literature in many areas. For master's level programs, a subscription to Ei's *CompendexPlus* is critical because it includes such a wide array of specialized literature. Access to additional specialized indexes, at least by a per-use fee, is also recommended. The *Science Citation Index* is useful for engineering but the expense is such that it isn't recommended unless the library would use it for multiple scientific disciplines. Furthermore, it does not index conference papers. If a program is focused on only one discipline, an alternative approach would be an index focusing on that discipline such as *Mechanical Engineering Abstracts* for Mechanical Engineering, or *INSPEC* for Electrical Engineering, with access to the interdisciplinary coverage on a per-use fee basis.

Monographs

Retrospective Purchases

Identifying books for the general collection is slightly more problematic. The guides to the literature in the previous section focus primarily on reference material; however some of the material could be considered for the circulating collection. Mildren's *Information Sources in Engineering* does list some titles that fit this description, and after 1993, some guides in the Engineering Literature Guides series include lists of important books. Hurt, Lord and Malinowsky list titles in categories such as handbooks, guides, tables or treatises that may also be considered for the general collection. In addition, two older sources give leads to some titles that may be useful for retrospective purchasing, and they may also be used to identify new editions of titles listed, or to identify works of the same authors. Those sources are:

> Powell, Russell. *Core List of Books and Journals in Science and Technology.* Phoenix: Oryx, 1987.

Association of College and Research Libraries. *Books for College Libraries*. 3rd ed. Chicago: American Library Association, 1988.

After identifying titles from these sources, there are several possible strategies to pursue to identify additional titles. When first selecting for the incipient programs at Boise State University, and before most of the new faculty were hired, the author sent a query out on ELDNET-L, the listserv sponsored by ELD, and COLLDEV-L, a listserv for collection development topics, asking what resources the members on the list found most useful for undergraduate students and the approaches taken to obtain materials for these students. Replies included specific series, publishers, societies, vendors, indexes, lists, review sources, and activities that different engineering librarians used to obtain these materials. Most of the comments here elaborate on suggestions received from that original posting.

* Develop a profile with a book vendor such as Blackwell North America or Yankee Book Peddler and ask for slips covering that profile for the previous year as well as the current year. Related to this is searching vendors' online files such as Blackwell's New Titles Online.
* Examine bibliographies of newer, high-quality undergraduate texts.
* Use review sources such as *Choice*, *SciTech Book News*, and *Aslib Book Guide* to identify reviews of recommended titles for the past few years.
* Ask faculty to identify titles of importance for the collection.

    Related questions for the faculty are (1) asking for relevant research topics and obtaining additional holdings in that area, (2) asking faculty to identify ten titles for each course they teach, or (3) asking for the faculty members' course syllabi or recommended reading lists and obtaining the books listed as recommended sources. This approach assumes there are faculty to ask, so if the program is really new these questions could be asked of faculty as they are hired for courses they had taught at previous colleges or universities.

* Search library holdings of schools that have similar programs.
* Identify occasional bibliographical references in the journal literature such as Ehrig's (1993) bibliography of Land Surveying and Crowell's (1999) discussion of bestsellers in the TP class.
* Identify books that have won awards such as those described by Smith (1998).

Using these various approaches, a large number of titles will be identified that when purchased, will give the collection a good foundation. As with any retrospective purchasing, numerous titles will be out-of-print. Engineering titles generally have small runs and are not frequently reprinted. There is an out-of-print market with some specialized services for technical materials. However, faculty input is recommended before purchasing many titles from this market.

Current Books/Ongoing Selection

The strategies for selecting current materials are very similar to other disciplines: approval plans, catalogs and review sources, and faculty suggestions. Each strategy is examined in further detail below.

- Set up a profile with an approval vendor for the topics, the audience, and the formats that will support the engineering program offerings. If the budget is small, define the profile for monographs coming in on approval very tightly, and use form selection to scan for additional relevant titles. Look over the approval vendor's selection of publishers to see which engineering societies are included. (See next paragraph for those that aren't.) For a selector's view of applying approval plans to engineering topics, see Franklin's 1989 article *Engineering Books on Approval: A Selector's Viewpoint.*
- Peruse review sources such as *Choice, SciTech Book News,* and *Aslib Book Guide.* Also scan magazines that feature new books such as *IEEE Spectrum, Civil Engineering,* and *Mechanical Engineering.* Many of these books will be coming through the approval plan, but occasionally publishers not included in the approval plan or titles that may be related but are not included in the subject profile will be found. For those engineering societies that aren't covered by the approval vendor, ask to receive the society's publications catalogs, or, if available, periodically review the society's new offerings on their web-based publications catalog.
- Ask faculty to send suggestions of titles that they consider appropriate to the collection. Send forms from the approval plan to faculty for their input. Ask faculty to prioritize their book orders–the use of a scale from 1-3, with 1 being very important and 3 being useful but not essential, has been successful at Boise State University.

These three strategies should result in an affordable collection that contains a good cross-section of the published books in the areas of emphasis. Periodically review the collection to identify gaps; fill in those gaps using strategies from the previous section on purchasing retrospective monographs.

Serials

Serials are critical in obtaining current news and reports on recent research. Collection funding ratios for serials to monographs in engineering run from 65/35% to 75/25%. Fahy (1990, 126) reported a ratio of 66.3/33.7% for the University of Arizona Library in 1988-89. Devin and Kellogg (1990) contend that it is reasonable to determine serial/ monograph ratios from serial use statistics found in citation studies; their review of the literature identifies four studies related to engineering with serials use results varying from 61.9% to 75.8%. As with other materials, there are selected titles that are interdisciplinary, and other titles that are discipline specific. As stated previously, engineering societies are key to the published literature, especially in serials. If the budget is limited, focus on a selection of titles from the major societies of the engineering disciplines covered at the institution. In the United States, this means societies such as the American Society for Civil Engineering (ASCE), American Society for Mechanical Engineering (ASME), Institution of Electrical and Electronics Engineers (IEEE), and the American Institute for Chemical Engineering (AIChE). If the budget is more substantial, consider purchasing journal packages from these organizations; terms will vary but society packages typically result in significant savings. Besides the major engineering societies, there are literally hundreds of engineering societies devoted to specialized disciplines; a good source to identify those that may publish useful material is the *International Directory of Engineering Societies and Related Organizations*, by the American Association of Engineering Societies. The standard *Encyclopedia of Associations* published by Gale will also provide information on engineering societies, including information about their publications. Many of the large societies publish their own material, but societies may also publish their material through a commercial publisher. There are also many key titles in engineering that are published by non-society publishers.

Periodicals

Engineering titles range from moderate prices to very expensive titles. The Periodical Price Survey 2000 published in *Library Journal* lists the average price for engineering titles as $1034.58 when using the list of journals covered in the Institute for Scientific Information (ISI) databases. By contrast, the average price for engineering titles is $323.44 when the Ebsco*host* Academic Search journal list is used (Ketcham-Van Orsdel 2000). The latter is likely to be more relevant to medium-sized university libraries. To identify those titles that will be critical for the collection, consult with faculty. If faculty members have not yet been hired, begin creating lists for them to review as they are hired. There are a several bibliographies that can help in this task, and some strategies for identifying appropriate titles.

- Katz, William A., Berry G. Richards, and Linda S. Katz. *Magazines for Libraries*. New York: Bowker, 2000. Published every 2-3 years.

    Titles for nine engineering disciplines are covered in the engineering and technology section, plus a listing of general engineering titles.

- Powell, Russell. *Core List of Books and Journals in Science and Technology*. Phoenix: Oryx, 1987.

    Although this title is dated, many of the serials are still published.

- Lord's, Malinowsky's and Chen's guides list some of the major periodicals.
- Review lists of periodicals covered by engineering indexes such as *Applied Science and Technology Index* and *CompendexPlus (Engineering Index)*.
- Ask each new faculty member for a list of the most critical journals in their field; ask them to consider the needs of undergraduates also since many of the titles they identify will be research-oriented.
- Subscribe to at least one news magazine for each discipline such as *Civil Engineering, Mechanical Engineering*, and *IEEE Spectrum*.

Conference Proceedings

Should conference proceedings be a part of the collection? This will depend on the makeup of the programs at the institution, and the faculty's recommendations. If the budget is small, focus on journals. However, the published account of conference proceedings becomes increasingly important as the degree level moves from bachelor to more advanced degrees. Topics are usually first published in conference proceedings. Later, articles may be published in the journal literature. So indexes covering conference proceedings are important to doctoral students and many master's students, and the corresponding conference proceedings are also important. Not every conference publishes the papers of the proceedings, but hundreds of established conferences do. Conference proceedings may be published by the conference, by the society sponsoring the conference, by a commercial publisher, or as a special issue in a journal. They can be annual, biennial, or have other variations. Published proceedings may be single-volume or multi-volume, and there may be series within a general conference proceedings. If conference proceedings are collected, here are some suggestions for identifying the publications:

- Ask the faculty to identify key conferences, and ask them about the importance of obtaining those proceedings for the collection.

- Review guides such as the Chen's, Mildren's, and the ELD Literature Guides Series that mention important conferences.
- Review the list of conferences indexed in *CompendexPlus*.
- Conference proceedings may be covered in the approval plan also, but to obtain all of the proceedings from a selected conference, use a serials vendor to set up a standing order, or set up a standing order directly with the publisher.

Other Serials

Yearbooks and Annual Reviews are also worth considering for ownership. These sources review the research of the past year and are usually excellent summaries. They are particularly helpful for students in advanced studies. Guides to the literature that include these materials include Chen, Mildren, and selected ELD guides.

Access

Serial literature decisions should be made in conjunction with decisions on services to obtain the articles or papers as needed–interlibrary loan and document delivery. Consider being conservative in the number of serials subscribed to, publicize the availability of access services, and see what kind of use there is. This strategy is better suited for faculty and graduate students than for undergraduate students, since the timeframe for undergraduate projects is usually very limited. These are areas where collaboration with faculty is essential to develop the collection most likely to succeed in meeting the needs of the students, staff and faculty.

At Boise State University, a pilot document delivery project was set up for faculty using *CompendexWeb*. Although there have been a few bumps in the road, this service has allowed faculty to quickly obtain material that Boise State University does not own. The "bumps" were mainly related to the business side of the operation and due to these limitations, students usually rely on interlibrary loan.

Other Material

In addition to the standard types of literature discussed up to now, engineers also refer to specialized materials that are useful in engineering research. These specialized materials include technical reports, patents, standards, and product catalogs. For a small to medium-sized program, it is unlikely that extensive collections of these materials will be needed. It is, however, important to be aware of their existence and the value that they have in engineering research.

Technical reports are the written results of research, usually with a great deal of detail. The reports may be written as a requirement of government-sponsored research, or for the record of a particular company, laboratory or agency. The National Technical Information Service (NTIS) has been the primary disseminator of technical reports sponsored by government funds. The United States Congress is currently debating the future of this organization so there may be changes in the current system. Some technical reports from companies are disseminated beyond the company, but many are not. Their existence is usually not known outside of the company until mentioned on a works cited page. Many can be difficult to track down.

Patents are another useful resource for engineers. In the United States, patents granted by the United States Government will be of primary interest but patents granted by the governments of other countries may also be of interest. Identify the patent depository library in the region, and discuss with the patent depository librarian how that library can help patrons access the United States patent information. Consider obtaining the *Index to the Patent Classification* and the *Definitions of Patent Classifications* from the Patent and Trademark Office in order to help patrons identify the appropriate patent classification. Full-text patents back to 1976 are also available on the Web for free.

Standards are the discipline's description of approved methods and procedures for activities such as testing, performance, construction, and terminology. Societies are absolutely key in this type of literature. There are thousands of standards. What is collected will be dependent on the programs the institution has, and the type of use that the faculty envision. Examples of well-known standards include the *Annual Book of ASTM* Standards (a multi-volume set), *ASME Boiler and Pressure Vessels Code*, and the *Uniform Building Code*.

Product catalogs are useful for identifying devices and materials to use when setting up research and design projects. The standard source for manufacturing companies is the *Thomas Register of American Manufacturing*, which includes a "yellow pages," a source book of companies, and catalogs of manufacturing companies who chose to include their catalog. This resource is also available on the Internet at: *http://www.thomasregister.com/*. There are a number of other directories that provide information about companies and the type of materials they supply. If the business collection is strong, these sources may already be present. In addition, many companies now have their product catalogs on the web, so finding sources to answer these needs is not so vexing as in the past.

Although each of these specialized types of material are important sources of engineering information, libraries supporting small to medium-sized programs will probably find it sufficient to purchase very selectively rather than to build extensive collections. Consult with the engineering faculty on the various types of literature to determine the best mix. Do identify sources that will help to obtain the documents at the time of need.

SERVICES

Although building the library collection is the major focus of this article, the impact on services is also important to consider. Interlibrary loan and document delivery services have been briefly mentioned. One should also address instruction and reference, and technical services.

Interlibrary loan for most materials requested for the students, staff or faculty of the engineering programs will be handled in the same way as other subject areas. For specialized material, it may be helpful to purchase the *Union List of Technical Reports, Standards and Patents in Engineering Libraries*, a publication, produced by ELD and sold by ASEE, that identifies libraries holding specific series of these categories. Document delivery may also be handled in the same way as other subject areas, but staff may want to check how the current services rate in providing engineering materials and look into services that specialize in technical literature if the current services are lacking.

As with other subject areas, activities introducing students and faculty to the resources available in the library, and helping them to use those resources, are a necessary component of support for new programs. This may entail staff training in the use and nature of engineering materials. Sources listed in the section on collecting reference material may be used to acquaint individuals with the technical literature. Attending conferences focusing on information services to engineering programs, such as the ASEE Annual Conference, may also be useful. A "train the trainer" model may be successful where one or two librarians become familiar with the holdings and the use of the material and then offer training sessions for those who staff the reference desk. These trainers may be the same people who will then offer instruction sessions to the students, staff and faculty. Although many engineering classes are laboratory based, there are usually some courses that are of a more general nature in which an introduction to the library can be incorporated. At Boise State University, opportunities for library sessions have occurred in the introduction to engineering class (freshman), engineering seminar (junior/ senior) and electrical engineering design project (senior). Once there are engineering graduate students on campus, a library orientation targeting those students is also envisioned. The focus of the instruction depends upon the class project or intent, but acquainting students with the literature of their field is valid even in lab-based curricula.

Technical services staff will probably also handle the new program's material in similar fashion to those of other subject areas. However, there are specialized distributors for out-of-print books and back issues of serials in the technical field, so identifying those would be of service. Setting up the appropriate approval plan will require some staff time, if this strategy is selected. Since there will be increased ordering, receiving, processing, and cataloging associated with the materials for the new program, the library administration may choose to request additional personnel to reflect the higher volume.

CONCLUSION

Building the library collection to support any new program demands time, consideration and funding. The main purpose of this article was to describe specific considerations and activities in building a collection to support small to medium-sized engineering programs and library funding levels. The literature supporting engineering is large and varied. A mix of standard selection practices using tools, vendors, and publishers covering the engineering literature and continued attention to assessment and updating collection development policies combined with collaboration with faculty will result in a collection ready to meet the needs of the engineering faculty, staff and students.

REFERENCES

Allen, Christine. 1998. Afterthoughts on collection development for new schools. *Book Report* 17(3): 8-9,11,91-92.

Crowell, Francoise. 1999. Bestsellers in chemical technology (subclass TP) 1998/99. *Against the Grain* 11(3): 52, 54.

Devin, Robin B. and Martha Kellogg. 1990. The serial/monograph ratio in research libraries: Budgeting in light of citation studies. *College and Research Libraries* 51: 46-54.

Ehrig, Ellen. 1993. Land surveying: A bibliography of sources. *RSR: Reference Services Review* 21(1): 65-70.

Fahy, Terry W. 1990. The need for a publishing universe (comparison of acquisitions intensity with total publisher output by subject; case of the University of Arizona). In *Acquisitions '90: Conference on Acquisitions, Budgets, and Collections, May 16 and 17, 1990, St. Louis, Missouri: proceedings.* Canfield, OH: Genaway.

Franklin, Hugh. 1989. Engineering books on approval: A selector's viewpoint. *Technicalities* 9(3): 12-13, 15.

Herring, Susan. 1992. The academic library materials budget and southeastern engineering programs. *The Southeastern Librarian* 42 (Spring): 11-13.

Johnson, William T. 1999. Environmental impact: A preliminary citation analysis of local faculty in a new academic program in environmental and human health applied to collection development at Texas Tech University Library. *LIBRES: Library and Information Science Research* 9(1): 1-19. Available from *http://aztec.lib.utk. edu/libres/libre9n1/toxcite.htm.*

Ketcham-Van Orsdel, Lee and Kathleen Born. 2000. Periodical price survey 2000: Pushing toward more affordable access. *Library Journal*, 15 April, 47-52.

Lanier, Don and Kathryn Carpenter. 1994. Enhanced services and resource sharing in support of new academic programs. *Journal of Academic Librarianship* 20: 15-18.

Linklater, Bill. 1988. Collection development in action. *Australian Academic and Research Libraries* 19 (Mar): 32-38.

Lowman, Shirley A. and Betty L. Blatz. 1988. Open your eyes before opening day: Providing collections for new community college centers. *Technicalities* 8(9): 10-11.

Myers, Marilyn. 1991. How the west was won: Building a collection in a new institution. *Library Acquisitions: Practice & Theory* 15: 39-43.

O'Gorman. Jack. 1999. Core collection: Engineering reference sources. *Booklist* 96: 732.

Pearson, Jeffrey W. 1999. Building a new undergraduate library collection. *College & Undergraduate Libraries* 6(1): 33-45.

Posey, Edwin D. 1987. Engineering. In *Selection of library materials in applied and interdisciplinary fields*. Chicago: American Library Association.

Powell, Jill. 1991. Beyond Compendex*Plus: A survey of selected core engineering compact disc databases. *CD-ROM Professional* 4(5): 55-68.

Powell, Jill H. 1993. Choosing engineering CD-ROM titles: Results of an Internet survey. *CD-ROM Professional* 6(6): 34-38.

Smith, Charles H. 1998. "With eyes on the prize": Identifying award-winning academic books in science and technology. *College & Undergraduate Libraries*, 5(2): 115-129.

Stevens, Norman D. 1991. Faculty start-up costs and library support. *College and Research Libraries News* 52: 491-493.

Woodberry, Evelyn. 1993. Issues in the establishment of a new serials collection. *Australian & New Zealand Journal of Serials Librarianship* 4(2): 69-76.

Yutani, Eiji. 1993. Developing new Japanese collections and services at the University of California, San Diego, 1988-1993. *Bulletin of the Association for Asian Studies* 101 (Dec): 75-8.

Resource Sharing in Engineering and Science Libraries

Charlotte A. Erdmann

SUMMARY. This paper examines the background and general principles of resource sharing in engineering and science libraries. It provides an understanding of the collections, electronic access, user expectations, and resource sharing necessary to meet the information needs of library clients. The author surveys interlibrary loan, document delivery services, and just in time purchases, and also presents case studies and bibliographic citations on these topics. Modern technical advancements and reciprocal, consortial, or contractual arrangements make it possible to deliver these services with faster turnaround times and lower costs. The paper also discusses specialized materials commonly used by engineers and scientists and recommended methods for obtaining these materials. *[Article copies available for a fee from The Haworth Document Delivery Service: 1-800-342-9678. E-mail address: <getinfo@haworthpressinc.com> Website: <http://www. HaworthPress.com> © 2001 by The Haworth Press, Inc. All rights reserved.]*

KEYWORDS. Resource sharing, interlibrary loan, document delivery, just in time purchases, science libraries, engineering libraries

Charlotte A. Erdmann is Assistant Engineering Librarian/Associate Professor of Library Science, Siegesmund Engineering Library, Potter Center, Purdue University, West Lafayette, IN 47907. She was a Fellowship Librarian at the U.S. Patent and Trademark Office in 1998-1999 and Information Services Librarian at South Dakota School of Mines and Technology from 1980-1984.

[Haworth co-indexing entry note]: "Resource Sharing in Engineering and Science Libraries." Erdmann, Charlotte A. Co-published simultaneously in *Science & Technology Libraries* (The Haworth Information Press, an imprint of The Haworth Press, Inc.) Vol. 19, No. 3/4, 2001, pp. 39-55; and: *Engineering Libraries: Building Collections and Delivering Services* (ed: Thomas W. Conkling, and Linda R. Musser) The Haworth Information Press, an imprint of The Haworth Press, Inc., 2001, pp. 39-55. Single or multiple copies of this article are available for a fee from The Haworth Document Delivery Service [1-800-342-9678, 9:00 a.m. - 5:00 p.m. (EST). E-mail address: getinfo@haworthpressinc.com].

39

OVERVIEW

The goal of libraries is to meet the information and research needs of their clients. An academic engineering library purchases books, subscribes to journals, acquires access to electronic indexes, and collects technical reports, standards, and patents. Clients become aware of the vastness of information available in print and on the Web and expect that any publication identified by an electronic index or printed source may be obtained through the library.

Since most library budgets do not permit the purchase of everything that a client needs, it is necessary to acquire materials through interlibrary loan, commercial document delivery services, and just in time purchases from bookstores, distributors, and publishers. These services are made possible by resource sharing, consortial, and contractual arrangements.

Lynden[1] sets the groundwork for his discussion of resource sharing by referring to the definition in the 1983 ALA Glossary. Resource sharing is:

> A term covering a variety of organizations and activities engaged in jointly by a group of libraries for the purposes of improving services and/or cutting costs. Resource sharing may be established by informal or formal agreement or by contact and may operate locally, nationally, or internationally. The resources shared may be: collections, bibliographic data, personnel, planning activities, etc. Formal organizations for resource sharing may be called bibliographic utilities, cooperative systems, consortia, networks, bibliographic service centers, etc.[2]

Lynden[3] also presents a variety of resource sharing mechanisms and describes "clues" of "what real resource sharing" includes. This is a simplified list:

- Requirements of users are most often satisfied.
- Materials are ready when the user needs them.
- Materials are readily available when the user is looking for them.
- Real resource sharing realizes savings.
- Real resource sharing reduces costs.
- Resources are increased locally.
- Resources are shared nationally.
- Resources of a monetary nature can be procured.
- Retention agreements are kept.
- Plans remain in place because top administrators look beyond immediate needs and support agreements.

Memberships in bibliographic utilities, cooperative systems, consortia, and networks allow libraries to deliver a variety of services that enhance access to information for library clients. Consortial and reciprocal agreements help libraries to provide resources and services at cost-effective rates. Libraries in consortia may also cooperate in the negotiation of licensing agreements for electronic resources and collaborate on purchasing expensive and rare materials.

Managing resources and collections during the shift from ownership to access has been a challenge for libraries and librarians. Many studies have been undertaken and many books and journal articles have been published about these issues. This paper will cover recent practice and research in interlibrary loan, document delivery, and just in time purchasing.

BACKGROUND

Today's engineering and science libraries contain collections of materials, competent staff, and a variety of services. Books and journal subscriptions are the major collections in academic libraries. Books contain state-of-the-art information but may quickly become dated. Patents, conference papers, technical reports, and journal articles contain much more timely information. Journals and electronic sources are the most expensive items that a library may buy or lease.

Clients typically expect Web access to the library catalog and electronic sources from their offices, laboratories, and homes. In some cases, these catalogs and electronic sources provide access to full-text images of electronic journal articles (e-journals), books (e-books), technical reports, and patents. Libraries usually pay extra for access to these images. Clients may request journal articles or books that are available locally through a campus photocopy service. Each academic or corporate setting has its own policies that determine who is eligible for this service. Some charge for this service while others provide the service free.

Whether the library's collection is large or small, it is usually not possible to own a collection that meets all clients' information needs. The exponential growth of information, new electronic services, new research areas, and rising materials costs compete for scarce resource dollars in all types of organizations. From talking with colleagues across the country and reading the library literature, it is apparent that most libraries have moved to an access model. Given the high costs of books and journals, it is no longer possible to rely entirely on purchasing materials just in case someone may read or view these materials someday.

Since most library budgets only permit the purchase of a limited amount of materials to support the teaching and research needs of organizations, it is necessary to use the shared resources of many libraries to meet clients' specialized needs. Chambers discusses many challenges that face libraries "before a total integrated document delivery system becomes a reality."[4]

> Cancellations and decreased purchasing power are causing research collections to look more and more alike as libraries are reduced to buying essential materials that support core programs. Vendors are in the business for a profit and what they make available for document delivery will be driven by the marketplace, not by the needs of individual scholars for esoteric, rare or foreign publications. A critical challenge for research libraries is to act collectively in a comprehensive and systematic manner to prevent the gaps of today from resulting in permanent loss of access for future generations of scholars and to ensure that unique literary manuscripts and archives–the primary materials that are the basis of research and scholarship–continue to be collected. This collective responsibility for our published heritage encompasses not just ownership, but also bibliographic control, preservation, archiving and delivery.[5]

With these challenges in mind, this article now examines the general principles involved in obtaining materials from other libraries, commercial services, and just in time services.

GENERAL PRINCIPLES

Materials that are not available locally may be acquired from other sources. The general principles of supplying materials as quickly as possible involve:

- Infrastructure of services, materials, computers, software, fax, and photocopiers. Consortia, reciprocal, or contractual agreements with suppliers. Deposit accounts may also be necessary.
- Knowledgeable and efficient staff members who identify, retrieve, and photocopy or loan materials.
- Willingness to use other libraries or commercial suppliers and pay supplier and copyright fees (if needed). Copyright fees are paid to the Copyright Clearance Center in the United States or similar centers in other countries.
- Library or commercial supplier with source (or ability to obtain source from another supplier) who agrees to supply.
- Efficient delivery system, e.g., fast–Ariel, fax, or courier; regular–U.S. mail or commercial delivery service, e.g., UPS.
- Time.

Methods that shorten the time from request to delivery are a priority. Depending on local policy, the library or the individual may pay for the service (libraries tend to finance part or all of the costs). Sometimes, individual libraries or consortia partners have agreements to fill loans and photocopies at no cost and give each other priority in processing and delivering materials.

INTERLIBRARY LOAN

Basics

When information is not accessible locally, loans of books, dissertations, and technical reports as well as photocopies of journal articles and conference papers are obtained from other libraries by a service commonly called interlibrary loan. The strengths of this approach are the wide variety of materials available from a large number of libraries, cooperative agreements among selected libraries that may provide loans or photocopies of materials at no cost, and electronic ordering features that send requests to a group of libraries. If the first library cannot supply a request, it is forwarded to the next library. The major weaknesses of interlibrary loan are the labor-intensive nature of the service, unavailability of materials, slow turnaround time, charging and payment of fees, and copyright tracking. The use of modern technologies is helping to overcome some of the disadvantages. Interested librarians may read Boucher's *Interlibrary Loan Practices Handbook*[6] to obtain an overview of borrowing, lending, and the relevant U.S. copyright law.

Requests in academic libraries are usually sent to other libraries (borrowing) and received from other libraries (lending) through one of the bibliographic utilities, e.g., Online Computer Library Center (OCLC), Research Libraries Information Network (RLIN), or Western Library Network (WLN). In December 1997, Jackson[7] reported on an Association of Research Libraries study of interlibrary loan/document delivery at 119 North American research and college libraries that indicated "on average, the unit cost to research libraries to borrow an item on interlibrary loan is $18.35, and the cost to lend an item is $9.48. Average borrowing turnaround time is 16 calendar days, the borrowing fill rate is 85%, and the lending fill rate is 58%." The study analyzed four performance measures: direct costs, fill rate, turnaround time, and user satisfaction. Characteristics of high performing borrowing and lending operations are identified. The author of this article suggests that readers investigate this study[8] in order to improve services in their libraries. The executive summary and frequently asked questions are enlightening.

Interlibrary loan services are encouraged to follow the U.S. National Commission on New Technological Uses of Copyrighted Works (CONTU) Guidelines. For periodicals or collective works, Boucher explains the fair use guideline, 1-1-5-5: "during one calendar year, for one periodical title, five articles can be copied from a title published in the last five years."[9] There are several alternatives when a library reaches the sixth copy of a periodical article or collected work. One of the common practices is to obtain a photocopy through ILL and "pay royalties to the copyright owner or the Copyright Clearance Center (CCC)."[10] Nonprofit, academic photocopying is covered by the 1-1-5-5 guideline. Fair use does not apply to corporate libraries so copyright fees must be paid on all copies. Ardito[11] has recently written about problems in obtaining copyright permissions since some publications are not registered with the Copyright Clearance Center.

Interlibrary loan offers a comprehensive program for obtaining all types of materials available at most libraries. Using bibliographic utilities and specialized sources, the service has access to the collections of thousands of libraries. That said, many specialized resources are not readily identified using online library catalogs and bibliographic utilities. *The Union List of Technical Reports, Standards and Patents in Engineering Libraries*[12] published by the Engineering Libraries Division of American Society for Engineering Education is a helpful source for locating holdings from academic engineering libraries for these specialized sources.

Conference Papers

The field of engineering has a number of societies that publish conference proceedings. Many libraries subscribe to the print, microfiche, or full-text publications of these societies. Some societies have specialized print or computerized indexes of their publications. Many indexes are free while a few are available only for subscribers. Conference papers may be indexed in INSPEC, COMPENDEX, or other databases available from online hosts or by subscription. The society's own index is usually more complete. Often, the societies offer document delivery services. A forthcoming article by the author of this article will outline more specifics of engineering society publications.

Since conferences proceedings can be difficult to decipher, skill and patience are needed to identify a conference in bibliographic utilities, e.g., OCLC and RLIN. The British Library also has created an *Index to Conference Proceedings* and includes conference records in its online catalog.[13] The author of this article occasionally uses the CONF (Conferences in Energy, Physics, Mathematics, and Chemistry)[14] database on STN International and Online-Katalog (TIBKAT)[15] at Universitaetsbibliothek Hannover to identify a conference that is not readily available on OCLC or RLIN.

Dissertations and Theses

Dissertations are another specialized source that engineers and scientists commonly use. Some U.S. libraries lend dissertations created at their institutions. Others must be purchased just in time from UMI Dissertation Services of

Bell and Howell Information and Learning or the university. Copies of most master's theses are obtained from the institution. The Center for Research Libraries[16] in Chicago is the prime collector and lender of more than 700,000 international dissertations for its member libraries. CRL also acquires dissertations on demand for its clientele.

Electronic theses and dissertations (ETDs) are a relatively new development. Virginia Tech is a leader in the National Digital Library of Theses and Dissertations (NDLTD). Tennant[17] and McMillan[18] report on a recent conference. Four educational institutions are now requiring submission of theses and dissertations electronically. Seventeen institutions have made 7,455 ETDs available. Many ETDs may be searched on Global Federated Search.[19] Many are available at no cost. UMI has also digitized more than 100,000 dissertations.

Access issues create roadblocks to acquiring some theses and dissertations. Due to funding issues, pending patents, or forthcoming publications, some dissertations or theses are not released immediately after the degree is granted. Thompson[20] studied access levels assigned to dissertations and theses at Virginia Tech. Concerns that putting ETDs on the Web "might jeopardize future publication of the material in books or refereed journals, might compromise pending patents, or might reveal too much concerning future research plans at Virginia Tech" may influence the use of restricted or withheld access levels. These concerns existed when dissertations or theses were printed on paper but to a lesser degree since these publications were not widely distributed or purchased.

Translations

Scientific and engineering publications are printed in many languages. Translations may be necessary for clients who do not understand the source language. Many translations are created on demand while others are published in cover-to-cover translation journals. Before paying an expert to translate a publication, it is recommended that one determine whether a translation is already available. One should not expect that a translation is available for every international language publication. Some indexing services may index only the original publication while others may index the original and the translation. For example, *Chemical Abstracts* indexes the original journal article and the first publication of a patent. Many translations are not indexed in traditional indexing and abstracting services.

Harman[21] of the British Library Document Supply Center (BLDSC) examined the acquisition of translations and specialized indexes and databases that provided access to more than 500,000 translations in 1988. The Translation Centre received on average 40,000 requests per year. Toot[22] of California Institute of Technology gives an excellent overview and instructions on the use of specialized translation indexes that have been published in the past as well as some databases and translation suppliers that are currently available. Canada Institute for Scientific and Technical Information (CISTI) is another organization that can supply a large number of translations. Current information from CISTI[23] and BLDSC[24] indicates that the libraries do not charge to check if a translation is available. Most translations are not listed in their catalogs. Dialog[25] makes available online, World Translations Index, 1979-1997 and suggests Delft Technical University, CISTI, and BLDSC for items found in the database. There are some national organizations that will do translations. The Japanese Information Center for Science and Technology (JISCT)[26] is offering a fee-based service for translations including manual and machine translation of Japanese to English.

Readers interested in machine translation are encouraged to pursue journal articles currently published on the topic, such as Balkin's recent discussion[27] of Babelfish that is used on Altavista and Lanza's description[28] of Globalink.

Case Studies

Fleck[29] describes new innovations at Michigan State University that are improving service. The articles gives detailed information about the user-initiated interface available through Innovative Interfaces system, Z39.50 software used to search library catalogs of Committee for Institutional Cooperation (CIC), and delivery tests using Ariel and Prospero to the client's desktop.

Turnaround time has varying levels of importance depending on the user. Some need requests quickly. Others can wait for a month or longer. Sellen[30] of SUNY at Albany describes the differences in turnaround time among four delivery systems. She uses two samples, i.e., four hundred requests in June, July, and August and two hundred requests in January, February, and March, from a private small undergraduate institution in Southern California. The delivery methods are: Ariel, courier (twice a week from libraries within ten miles), mail, and fax. The results show mean turnaround time (in days) for the following: Ariel 3.65 days [107 requests]; Courier 2.52 days [310 requests]; Mail 9.65 days [169 requests]; and Fax 2.94 days [16 requests]. The average for all four methods was 4.79 days. This was a significant improvement over Bud's study[31] in 1986 that showed the average to be 13.76 days and Jackson's study[32] in 1997 that averaged 16 days for research libraries.

Douglas and Roth[33] of California Institute of Technology discuss a locally developed user-initiated request system for on-campus delivery and interlibrary loan services. In 1989, Caltech staff initially loaded ISI Scisearch tapes for 1,800 journals owned by Caltech on TOC/DOC (Tables of Contents/Document Delivery). The service was expanded to 2,500 journals including some journals that are not available on campus. Caltech also provides a cost recovery photocopy service for campus clients with a 24/48-hour turnaround time. "Nearly 60% of the TOC/DOC photocopy requests came from 100 journals in 1991." Many journals were not used at all. Caltech studied use statistics and cancelled some low-use journals later. Many requests for materials not owned by the Institute are supplied through G4 (Group of 4) consortium agreement with University of California, Los Angeles. UCLA's Orion Express and UPS next day delivery "expedite interlibrary loan requests for books and journal articles." Caltech is currently using ISI's Web of Science and using some document delivery services.[34]

Colorado State University's successful system is an example of a large-scale user-initiated request program on the libraries' Web site. The use of modern technology and cooperative library partners made it possible for Colorado State to deliver services following a flood that destroyed much of the library collection. Delaney[35] describes the services and procedures needed to supply an enormous number of publications. Staff used Clio and many Ariel workstations to supply requests.

Siddiqui[36] examines one year of loan and photocopy requests at the King Fahd University of Petroleum and Minerals Library. The primary users are the faculty and graduate students. British Library Document Supply Center and the Bibliotheca TU Delft (The Netherlands) fill 76% of the 1,027 requests. Deposit accounts are active with frequently used library suppliers.

DOCUMENT DELIVERY SERVICES

Basics

A library may also obtain photocopies of journal articles or other publications from commercial document delivery services or fee-based library services. The strengths of this approach include lowering labor costs for the requesting library, providing faster delivery than interlibrary loan, and payment of copyright fees. Some of the weaknesses may be the use of multiple services, unexpected costs, and cancelled orders. Some delivery services do obtain materials from other organizations and suppliers for an added fee. The requestor must be willing to pay the costs. Typically, one service cannot supply all needed materials.

Ward[37] who is head of Access Services at Purdue University, offers advice for choosing suppliers for outsourcing document delivery requests. She discusses different types of suppliers, including general and specialized commercial document delivery services, publishers, database providers, national library collections, fee-based information services, and information brokers and criteria needed to choose suppliers. Some of the factors involved are good online access and database quality, knowledge of costs before ordering and average turnaround time, several ordering and delivery options, helpful customer service, and correct and timely invoices. Ward also suggests testing services for six months to compare services before choosing a service(s) as well as periodically evaluating existing services. Most document suppliers specialize in a subject area or a collection of materials. They can also obtain a source from another supplier. Simpson[38] identifies commercial suppliers and library-based services, and maintains a list on a Web site at the National Institute of Health. Online databases may also direct users to selected document suppliers.

When using document delivery services, Copyright Clearance Center (U.S.) fees must be paid for each journal article or publication. The document supplier may charge a base fee, copyright fee, and delivery charge. Some suppliers negotiate with publishers separately and may offer lower copyright rates.[39,40] Fees may vary among suppliers as a result.

Many libraries have successfully used document suppliers. Jackson and Croneis[41] compiled a SPEC survey, "Uses of Commercial Document Delivery Services" for ARL in 1994. For a specific request, libraries chose the supplier for speed of delivery and ease of ordering. Most respondents indicate that libraries are increasing the use of document suppliers. At that time, the most frequently used services were: Canada Institute for Scientific and Technical Information (CISTI), The Genuine Article (ISI), Carl Uncover, British Library Document Supply Center (BLDSC), Dialog Information Service, and UMI Article Clearinghouse.

Another possible source for a hard-to-find publication that is identified from a print index or its online counterpart is the publisher or database producer. Many maintain fee-based services or have contracts with document delivery services. Many producers have Web sites and also maintain links from online hosts. For example, ASM International[42] is one of the producers that is responsible for the creation of METADEX and maintains a library and document delivery service. JICST indexes Japanese publications for JICST-E (available on Dialog and STN). The JICST Library[43] provides a document delivery service that is available directly or through STN.

Commercial document delivery services and fee-based library services provide similar services. These services are changing constantly. Several are no longer doing document delivery. Ebsco, UMI and Ei Engineering Information no longer deliver journal articles and conference papers. Linda Hall Library[44] is now the supplier for Ei Engineering Information, AIAA Dispatch, and API Encompass and also contains a large collection of historic materials, society publications, and journals.[45] Most contents of the former Engineering Societies Library in New York are housed at the Linda Hall Library. Ward and Dugan[46] describe an academic fee-based information service at Purdue University that accepts requests from corporate, industrial, and professional clients. Many large universities have similar services.

Case Studies

Since the budget crunches of early 1990s, many libraries have cancelled journal subscriptions. In order to meet their clients' needs, some libraries have funded user-initiated document delivery services, such as Carl UnCover. This company offers quick turnaround facsimile delivery of journal articles from 18,000 journals with 8 million articles. This service can block journal titles that a library owns. In most cases, the cost of subscribing to cancelled journals is much greater than the requests for individual journal articles and copyright fees from document delivery services. Beam[47] documents the use of Carl UnCover at Colorado State and also highlights universities using UnCover in her literature review.

Widdicombe[48] describes a just in time interlibrary loan, document delivery, and purchase program at the S. C. Williams Library at Stevens Institute of Technology. Many special libraries in corporations are much like Stevens Institute of Technology. The library discontinued all subscriptions for journals in 1993. Stevens clients are much more interested in the journal article than the journal itself. Due to budget reductions, Widdicombe shifted acquisitions money to electronic sources and paid for interlibrary loan and document delivery services to supply the materials quickly. Carl Uncover and OCLC FirstSearch are some of the services that Stevens uses.

Clement[49] investigates suppliers in a pilot project at Pennsylvania State University in 1993-1994 for requests from the science branches (Earth & Mineral Sciences, Engineering, Mathematics, and Physical Sciences). Two hundred thirty-seven requests were ordered from six suppliers. "The team concluded that purchasing articles through commercial suppliers was a viable means of extending several areas of the collection in a timely and cost-effect manner."

Chrizastowski and Anthes[50] report on a 6 1/2 month pilot study of Chemical Abstracts Document Delivery Service in a decentralized Chemistry Library at the University of Illinois paid by gift funds. A goal of the library is to provide 90% of the user's needs in the on-site library. Rising serial costs have required two serial cuts in 1988 and 1993. The use of a document delivery service is a way to obtain the rest of the sources needed by the library's clients. An analysis of the journal titles requested from Chemical Abstracts shows 136 unique titles. "The majority of journal titles requested were never owned by the UIUC (111 or 82%). Titles that were once owned but cancelled represented 25 titles, or 18% of the journals." Eighty percent of the cancelled titles had only one request. The authors recommend continuing the use of document delivery services and suggest using more than one service since 15% of the requests could not be supplied and were forwarded to the campus interlibrary loan.

McFarland[51] discusses a comparison of document delivery services for requests in engineering and science in 1992. He requested twenty articles from nine services and compared costs, turnaround times, cost-efficiency, fill-rates, reliability, and vendor responsiveness. McFarland sees one of the advantages of using document delivery services is that the service handles "copyright compliance and payments to publishers." It is the library's responsibility to understand what is paid by the service. It could be interesting to redo McFarland's comparison using today's document delivery services and modern technologies, including Ariel and fax.

JUST IN TIME PURCHASES

Basics

Libraries are using just in time purchases of newer books, dissertations, technical reports, and standards to acquire materials quickly. Obtaining loans of these materials may sometimes be difficult. Libraries make purchases using online bookstores, electronic orders, fax orders, and phone orders. The use of credit cards and deposit accounts also speeds the process. The strengths of just in time purchases consist of relatively quick turnaround time and the addition of needed materials to the collection for the future client. The weaknesses are that some attempts to purchase materials may be labor-intensive and time consuming. Some materials may be unobtainable.

Perdue and Van Fleet[52] at Bucknell University reported on the purchase of books requested by users of interlibrary loan. The rationale for purchasing the ILL books from the library's collection development funds is that requested books can be expected to circulate. A study of several years' purchases indicated that the ILL requested books were used more frequently than other purchased books. Purdue University's Acquisitions Department is currently experimenting with the purchase of recently published books that are requested on interlibrary loan. Interlibrary loan departments and departmental libraries may also regularly purchase dissertations, patents, standards, and technical reports.

Grey Literature and Reports

Suppliers of grey literature make available a variety of publications not normally published by commercial publishers. These may include research, technical, and government reports as well as preprints, committee reports, working papers, conference papers, translations, and dissertations. This literature may be obtained from a distributor, government service, library, or creator. Some suppliers specialize in one type of grey literature, e.g., dissertations, standards, or patents. Others are multi-faceted suppliers. The format of the document may be paper, microform, CD-ROM, or full-text image. Selected reports are available on the Web for free or fee. Some reports have security restrictions that make them more difficult to obtain. Deposit accounts are frequently used for these purchases. This makes it easier to obtain reports quickly since no purchase order or check is created.

Typical U.S. government suppliers are the following: (a) National Technical Information Service (NTIS) which supplies reports from a variety of government agencies, e.g., Department of Transportation and Environmental Protection Agency; (b) National Aeronautics and Space Administration (NASA) which supplies NASA, NACA, and AGARD publications; (c) Defense Technical Information Center (DTIC); and (d) Department of Energy (DOE). NASA and DTIC give specialized services and access to their contractors. Librarians are encouraged to talk to the university or company contracts office and determine if the organization has any current contracts with these agencies that may qualify library clients for these services. It is also recommended that library staff have access to the NTIS and DOE Energy databases using an online host or subscription. These databases are useful for identifying many government reports. Recent NTIS reports are available from their Web site. DTIC and NASA also have many indexes and reports available on their Web sites. Chawla[53] of the University of Maryland Libraries has created *The Virtual Technical Reports Center* that is very useful for locating reports as well as "preprints, reprints, dissertations, theses, and research reports of all kinds." Many of these reports are full-text.

European research, technical and government reports are among the records contained in a multi-disciplinary database called SIGLE[54] (System for Information on Grey Literature in Europe) that is produced by EAGLE, a consortium of leading libraries in Europe. The database also includes preprints, committee reports, working papers, conference papers, translations, dissertations, and market surveys. The items in this database can be ordered from the source library identified in each record. The database contained over 565,000 records from 1980-date and is available through online hosts.

Dissertations

Most U.S. dissertations are easily purchased from UMI Dissertation Services[55] provided that the university requires participation and access limits are unrestricted. Many universities eventually release restricted, confidential, and withheld dissertations. UMI is making available more than 100,000 dissertations in PDF format. A library may charge the client for a dissertation. It then becomes the property of the client. UMI also allows purchase by credit card so that a client may purchase a dissertation directly.

Some educational institutions do not make their dissertations available through UMI. The dissertation is usually obtained from the institution's library. Selected universities make dissertations available on the Web.

Patents

U.S. patents are readily available from many sources. Libraries in all fifty states and Puerto Rico are members of the Patent and Trademark Depository Library Program from the U.S. Patent and Trademark Office. At the present time, most libraries have collections of microfilm and CD/DVD-ROM patents. Over the next few years, libraries will receive a complete set of all U.S. patents on DVD-ROM. Staff at the United States Patent and Trademark Office[56] recently loaded TIFE images of U.S. patents issued since 1790 on the Office Web site. The European Patent Office[57] maintains a Web site with 30 million patents in PDF format. Both USPTO and EPO sites permit printing one page at a time. To save staff time, it may be necessary to find suppliers for long patents. Fitzgerald[58] writes a short article that describes several more fee and free sources, including patents and legal information, e.g., Micropatent and Delphion Intellectual Property Network sponsored by IBM. Many patent and trademark depository libraries and document delivery services also supply patents.

Standards

Engineering researchers and practitioners frequently use standards and specifications. Loans of these materials may be limited. Fee-based full-text standards are available for many organizations by subscription, e.g., Informational Handling Service. Some full-text government specifications are available on the Web, e.g., DTIC. For others, selected standards may be ordered directly from the organization that created the standard or from a standards supplier, e.g., Global Engineering Documents (United States) or ILI (United Kingdom). Deposit accounts may also be set up with standards suppliers.

CONCLUSION

Resource sharing expands the capabilities of engineering and science libraries. Future access to information is influenced by continued access to critical collections of materials, modern technologies, and skilled staff members. Interlibrary loan, document delivery, and just in time purchases are three approaches that libraries use to acquire quality information for clients in a timely manner.

REFERENCES

1. Frederick C. Lynden, "Will Electronic Information Finally Result in Real Resource Sharing?" *Journal of Library Administration* 24, no.1/2 (1996): 47-72.

2. Young Headstall, "Resource Sharing" in *The ALA Glossary of Library and Information Science* (Chicago: American Library Association, 1983).

3. Lynden, 47-72.

4. Joan Chambers, "Determining the Cost of an Interlibrary Loan in North American Research Libraries: Initial Study," *62nd IFLA General Conference–Conference Proceedings*–August 25-31, 1996. 22 November 2000. <http://ifla.org/IV/ifla62/62-chamj.htm>.

5. Ibid.

6. Virginia Boucher, *Interlibrary Loan Practices Handbook*, 2nd ed. (Chicago: American Library Association, 1997).

7. Mary E. Jackson, *Measuring the Performance of Interlibrary Loan and Document Delivery Services.* 12/97. 27 November 2000. <http://arl.org/access/illdd/illdd-measperf9712.shtml>.

8. *Access & Technology Program/ILL/DD Performance Measures Study.* 4/27/1999. 27 November 2000. <http://www.arl.org/access/illdd/illdd.shtml>.

9. Boucher, *Interlibrary Loan Practices Handbook*, 68-76.

10. Ibid., 73.

11. Stephanie C. Ardito, "One More Barrier to Compliance: Publications Not Registered with the CCC," *Online* 24 (Jan/Feb 2000): 67-70.

12. Mary C. Schlembach, *Union List of Technical Reports, Standards and Patents in Engineering Libraries*, 4th ed. Engineering Literature Guides; no. 26 (Washington, D.C. : American Society for Engineering Education, 1999).

13. The British Library, Document Supply Centre, *Grey Literature–General. Factsheet C4.* 20 November 2000. <http://www.bl.uk/services/bsds/dsc>.

14. *ONF. STN Database Summary Sheet.* 9/2000. 15 January 2000. <http://www.cas.org./ONLINE/DBSS/confss.html>.

15. *Online-Katalog der UB/TIB Hannover.* 27 November 2000. <http://www.hobsy.de/cgi-bin/nph-wwwredir/sun31.tib.uni-hannover.de:53229/>.

16. The Center for Research Libraries, *International Doctoral Dissertations.* 1998. 22 October 2000. <http://wwwcrl.uchicago.edu/info/docdiss.htm>.

17. Roy Tennant, "Accessing Electronic Theses: Progress?" *Library Journal* 125, no.9 (May 15, 2000): 30-33.

18. Gail McMillan, "Managing Electronic Theses and Dissertations: the Third International Symposium," *College and Research Libraries News* 61, no.5 (May 2000): 413-414.

19. *Global Federated Search: Global Federated Searcher.* 10/5/2000. 14 November 2000. <http://jin.dis.vt.edu/fedsearch/>.

20. Larry A. Thompson, "Electronic Theses and Dissertations at Virginia Tech: A Question of Access, Session 1441" in *Proceedings, 2000 ASEE Annual Conference & Exposition, St. Louis, Missouri, June 18-21, 2000*, 24 pages <computer file>.

21. Maya Harman, "The British Library Document Supply Center as a Translations Centre," *Interlending and Document Supply* 16, no.1 (1988): 17-20.

22. Louisa Toot, *Translations Resources in the Caltech Libraries*. 4/26/2000. 30 October 2000. <http://library.caltech.edu/collections/guides/translation_guide/default.htm>.

23. CISTI, *The Library and Document Delivery: About the Collections, Scientific Translations*. 30 October 2000. <http://www.cisti.nrc.ca/cisti/irm/trans_e.shtml>.

24. Document Supply Center, *Catalogues and Collections: Translations*. 30 October 2000. <http://www.bl.uk/services/bsds/dsc/translations.html>.

25. Dialog. *World Translations Index (1979-1997)*. 3/2/1998. 30 October 2000. <http://library.dialog.com/bluesheets/html/bl0295.html>.

26. Japan Science and Technology Corporation, *Translation Services: JICST Provides Two Types of Translation Services*. 21 November 2000. <http://pr.jst.go.jp/EN/ServiceGuide/trans.html>.

27. Ruth Balkin, "Babelfish: AltaVista's Automatic Translation Program," *Database* 22, no.2 (April/May 1999): 56-7.

28. Sheri R. Lanza, "A Cure for Web Translation Blues: Globalink," *Database* 21, no.5 (October/November 1998): 57-59.

29. Nancy W. Fleck, "Interlibrary Loan–a New Frontier!" *Library Hi-Tech* 18, no.2 (2000): 172-176.

30. Mary Seller, "Turnaround Time and Journal Article Delivery: A Study of Four Delivery Systems," *Journal of Interlibrary Loan, Document Delivery & Information Supply* 9, no.4 (1999): 65-72.

31. John Budd, "Interlibrary Loan Service: A Study of Turnaround Time," *RQ* 26 (Fall 1986): 75-80.

32. Jackson, *Measuring the Performance of Interlibrary Loan and Document Delivery Services*. 12/97. 27 November 2000. <http://arl.org/access/illdd/illdd-measperf9712.shtml>.

33. Kimberly Douglas and Dana L. Roth, "TOC/DOC 'It Has Changed the Way I Do Science,'" *Science & Technology Libraries* 16, no.3/4 (1997): 131-145.

34. CLS Web Team, Caltech Library Systems, *Web of Science*. 6/20/2000. 23 October 2000. <http://library.caltech.edu/scisearch/default.htm>.

35. Thomas Delaney, "The Day It Rained in Fort Collins, Colorado," *Journal of Interlibrary Loan, Document Delivery & Information Supply* 8, no.4 (1998): 59-70.

36. Moid A. Siddiqui, "A Statistical Study of Interlibrary Loan Use at a Science & Engineering Academic Library," *Library Resources & Technical Services* 43, no.4 (October 1999): 233-46.

37. Suzanne M. Ward, "Document Delivery: Evaluating the Options," *Computers in Libraries* 17 (October 1997): 26, 28-30.

38. Jean Simpson, *Document Delivery Suppliers*. 8/15/2000. 7 September 2000. <http://www.nnlm.nlm.nih.gov/pnr/docsupp/AAA-FRAME-info.html>.

39. Paula Eiblum and Stephanie C. Ardito, "Royalty Fees Part II: Copyright and Clearinghouses," *Online* 22, no.2 (May/June 1998): 51-56.

40. Stephanie C. Ardito and Paula Eiblum, "Royalty Fee Part III: Copyright and Clearinghouses–Survey Results," *Online* 22, no.4 (July/Aug 1998): 86-88, 90.

41. Mary E. Jackson and Karen Croneis, *Uses of Document Delivery Services*, SPEC Kit 204 (Washington: Association of Research Libraries, 1994).

42. ASM International. *News and Resources: ASM Library*. 4 January 2001. <http://www.asminternational.org/content/NewsandResources/Library/library.htm>.

43. JST, Information Center for Science and Technology, *Document Delivery Service, JICST Library*. 21 November 2000. <http://pr.jst.go.jp/EN/ServiceGuide/ext-serv.html>.

44. *Linda Hall Library: Science, Engineering, and Technology.* 23 October 2000. <http://www.lindahall.org/>.

45. Bruce Bradley, "A Library of First Resort for Science, Engineering and Technology: the Linda Hall Library" in *Resources Management, Proceedings of the 16th Biennial IATUL Conference, Enschede, The Netherlands, June 5-9, 1995*, edited by Gerard A.J.S. van Marle and Elin Tornudd. Enschede: Published for IATUL by the University Library of Twente, 1995.

46. Suzanne M. Ward and Mary Dugan, "Document Delivery in Academic Fee-Based Services," *The Reference Librarian*, no.63 (1999): 73-81.

47. Joan Beam, "Document Delivery via UnCover: Analysis of a Subsidized Service," *Serials Review* 23, no.4 (Winter 1997): 1-14.

48. Richard P. Widdicombe. "Eliminating All Journal Subscriptions Has Freed Our Customers to Seek the Information They Really Want and Need: The Result–More Access, Not Less," *Science & Technology Libraries* 14, no.1 (1993): 3-13.

49. Elaine Clement, "A Pilot Project to Investigate Commercial Document Suppliers." *Library Acquisitions: Practice & Theory* 20, no.2 (1996): 137-146.

50. Tina E. Chrzastowski and Mary A. Anthes, "Seeking the 99% Chemistry Library: Extending the Serials Collection Through the Use of Decentralized Document Delivery," *Library Acquisitions: Practice & Theory* 19, no.2 (1995): 141-152.

51. Robert T. McFarland, "A Comparison of Science Related Document Delivery Services," *Science & Technology Libraries* 13, no.1 (Fall 1992): 115-135.

52. Jennifer Perdue and James A. Van Fleet, "'Borrow or Buy?' Cost-Effective Delivery of Monographs," *Journal of Interlibrary Loan, Document Delivery & Information Supply* 9, no.4 (1999): 19-28.

53. Gloria Lyles Chawla, *The Virtual Technical Reports Center: EPrints, Preprints, & Technical Reports on the Web.* 8/29/2000. 12 September 2000. <http://www.lib.umd.edu/UMCP/ENGIN/TechReports/Virtual-TechReports.html>.

54. *The European Association for Grey Literature Exploitation.* 28 November 2000. <http://www.kb.nl/infolev/eagle/frames.htm>.

55. Bell + Howell Learning and Information. *UMIDissertation Services.* 28 November 2000. <http://www.umi.com/hp/Support/DServices/>.

56. U.S. Patent and Trademark Office, *Full-Text Database.* 11/15/2000. 28 November 2000. <http://www.uspto.gov/patft/index.html>.

57. The European Patent Office, *Europe's Network of Patent Databases.* 11 November 2000. <http://ep.espacenet.com/>.

58. Marc C. Fitzgerald, "A Patent Miner's Story." *Chemical Innovation* 30, no. 54-55 (November 2000) <http://pubs.acs.org/subscribe/journals/ci/30/i11/html/11inet.html>.

Grey Literature in Engineering

Larry A. Thompson

SUMMARY. Grey literature, such as meetings papers, technical reports, manufacturers' catalogs, and industry standards, are critical to engineering research. Unfortunately, this literature is often not readily available, not cited properly, and not indexed comprehensively. All of these factors make it difficult to access. This paper discusses the major areas of engineering grey literature and suggests ways that it can be acquired and used. *[Article copies available for a fee from The Haworth Document Delivery Service: 1-800-342-9678. E-mail address: <getinfo@haworthpressinc. com> Website: <http://www.HaworthPress.com> © 2001 by The Haworth Press, Inc. All rights reserved.]*

KEYWORDS. Grey literature, gray literature, engineering, proceedings, standards

INTRODUCTION

Grey literature covers the gamut of publications, from one page datasheets to multi-volume technical reports, from freely available government docu-

Larry A. Thompson, BS, MLS, is Assistant Professor and Engineering Librarian, Virginia Polytechnic Institute and State University, Blacksburg, VA 24062-9001 (E-mail: larryt@vt.edu).

[Haworth co-indexing entry note]: "Grey Literature in Engineering." Thompson, Larry A. Co-published simultaneously in *Science & Technology Libraries* (The Haworth Information Press, an imprint of The Haworth Press, Inc.) Vol. 19. No. 3/4. 2001. pp. 57-73; and: *Engineering Libraries: Building Collections and Delivering Services* (ed: Thomas W. Conkling, and Linda R. Musser) The Haworth Information Press, an imprint of The Haworth Press. Inc.. 2001, pp. 57-73. Single or multiple copies of this article are available for a fee from The Haworth Document Delivery Service [1-800-342-9678, 9:00 a.m. - 5:00 p.m. (EST). E-mail address: getinfo@haworthpressinc.com].

ments to reports costing tens of thousands of dollars, from meetings papers available only in paper format to standards available through download from the WWW. If you can imagine a publication format or mode of delivery, there will most likely be a grey literature publisher using it.

Added to this variety is the uncertain flow of change for grey literature. With mainstream commercial journals it's reasonable to expect that most will produce web versions in order to be competitive and meet the needs of their subscribers. With grey literature, such wholesale movement is not so easily predicted.

The very nature of grey literature, including its diverse sources and variety of formats, guarantees that its migration to the electronic age will be full of fits and starts. Thus, this paper only gives a snapshot of the current situation, one that could (and most likely will) change tomorrow.

THE DEFINITION OF GREY LITERATURE

Grey literature has been defined in various ways. The Grey Literature Network Service states that grey literature is "That which is produced on all levels of government, academics, business and industry in print and electronic formats, not controlled by commercial publishers."[1] This definition is sometimes referred to as *The Luxembourg Convention on GL*, because it was formulated at the Third International Conference on Grey Literature, GL'97, in that city.

Auger discusses the definition of grey literature at length, stating that it includes "reports, technical notes and specifications, conference proceedings and preprints, supplementary publications and data compilations, trade literature, and so on."[2] In addition to having uncertain availability, he states that grey literature is characterized by "poor bibliographic information and control, non-professional layout and format, and low print runs."[3] He also notes that some have defined grey literature as "literature which is not readily available through normal book-selling channels, and therefore difficult to identify and obtain."[4] He questions this latter definition because some items, although not readily available through booksellers, are easily available from other sources.

This paper will adopt the broad definition of the Luxembourg Convention. However, not all types of grey literature covered in this paper will be discussed at equal length. As Auger has noted, some literature that is considered grey is actually quite well indexed and easy to obtain. Therefore, as literature types are discussed some will require more detail than others.

TECHNICAL REPORTS–U.S. GOVERNMENT

General Observations

In some cases the distribution of government sponsored technical reports has changed over the years. Many U.S. government report series were originally under the jurisdiction of the Government Printing Office (GPO), given a Superintendent of Documents (SuDoc) number, and distributed freely as part of the Federal Depository Library Program (FDLP). During the 1980s and 1990s some of these report series were placed under the jurisdiction of the National Technical Information Service (NTIS). Therefore, they were no longer freely available through the FDLP, but were available only by purchase from the NTIS. As a result, libraries often have incomplete runs of series, terminating when the status changed to an NTIS distributed document.

In part because of this switch in jurisdiction from the GPO to the NTIS, the NTIS has a significant collection of documents related to engineering. Its collection has nearly three million publications and it adds tens of thousands of publications each year. Among the agencies of interest to engineering that contribute to the NTIS are the Department of Transportation (DOT) and the Environmental Protection Agency (EPA).

With respect to the NTIS, it should be noted that for the most part NTIS distributes reports, but it does not produce them. Documents that are referred to as "NTIS technical reports" are actually generated by other agencies, (e.g., EPA, DOT), and sent to NTIS for indexing, archiving, and distribution. Because NTIS is only a distributor, when a government report series changes from GPO to NTIS, it is not the content or producer that changes, but only the distributor.

Over time the federal government has also changed distribution formats. Most reports were originally distributed in paper. During later years many reports were issued in microfiche format, or sometimes were offered in a choice of the two formats. In recent years, more government documents have been distributed on CD-ROM and it has been announced that some documents will be issued in DVD format. The WWW is also becoming increasingly important, with much information readily available online.

As government agencies change the distribution methods and formats of their reports, the strategy for accessing the reports must also change. The following sections look at several government agencies and the access methodology for each.

Environmental Protection Agency

Many EPA reports have followed the pattern noted above, changing from GPO to NTIS. Thus, an early report in a series may be found in the GPO database and have a SuDoc number. Later reports in the same series may not have a

SuDoc number and may be found only in the NTIS database because they were distributed by the NTIS. Because NTIS distributed the reports, and because the NTIS database is a major resource for searching them, many libraries have shelved NTIS reports according to the NTIS accession number. This is especially true with microfiche. In order to locate the report in the collection it's necessary to search the NTIS database for the document and retrieve the NTIS accession number.

The EPA reports are increasingly available on the WWW, and the EPA provides several ways to search its WWW site and/or publications. The EPA National Publications Catalog (http://www.epa.gov/ncepihom/catalog.html) allows searches for publications that are available through the National Service Center for Environmental Publications, located in Cincinnati, OH. The National Environmental Publications Internet Site (NEPIS, http://www.epa.gov/ncepihom/nepishom/), searches EPA for documents that are available online. The search engine is very basic and not very user-friendly. A much better search engine with more search options was found linked off the EPA home page (http://www.epa.gov/epahome/pubsearch.html).

When a search was done on the title *25 Years of the Safe Drinking Water Act: History and Trends*, using the tools listed above, the results were not encouraging. Neither the EPA National Publications Catalog nor NEPIS provided any information. Using the search engine linked off the EPA home page did provide ordering information for the document. Unfortunately, none of the search engines indicated that the entire full text document is available online. The full text of the document was eventually found by browsing the site, not a very efficient way to do research.

This example from the EPA illustrates the problems that can be present in the web environment. Search engines do not always retrieve all information that is available online and several different search engines may be available to search the site and/or the agency.

NACA and NASA

The National Advisory Committee for Aeronautics (NACA) was the precursor to the National Aeronautics and Space Administration (NASA). NACA (1915-1958) began in the midst of World War I, while NASA (1958-present) was created following the Sputnik crisis. Both have produced huge amounts of literature. In contrast to much of the sci/tech literature, NACA/NASA literature from the 1950s and 60s is still heavily used by researchers and students. Basic research done during that time when the United States was gearing up its space program is still valuable.

NASA, like most information providers, is continually updating its information services to reflect the influence of the WWW. As with the EPA example, users must navigate a variety of interfaces and search engines to locate full text documents. A general overview is given in order to facilitate information retrieval.

The NASA Scientific and Technical Information (STI) Program collects information from U.S. and international sources and places bibliographic records for these items in the STI database. General public access to the STI database through the Center for Aerospace Information Technical Reports Server (CASI TRS, http://www.sti.nasa.gov/RECONselect.html), encompasses only a very small portion of the entire material in the database. The Scientific and Technical Aerospace Reports (STAR, http://www.sti.nasa.gov/Pubs/star/Star.html) is an electronic abstract journal. It is published on the WWW and contains the same publicly available information that is available from the NASA STI database as searched through the CASI TRS. The NASA Technical Reports Server (NASA TRS, http://techreports.larc.nasa.gov/cgi-bin/NTRS) provides access to several different databases in different geographical locations.

Department of Energy

The Department of Energy (DOE) is becoming a leader among government agencies in providing access to grey literature on the WWW. Their Office of Scientific and Technical Information (OSTI) has created a web page (http://www.osti.gov) that provides access to three types of material: peer-reviewed journals, grey literature, and preprints. The PubSCIENCE database indexes peer-reviewed journal literature, but is not as comprehensive as databases such as Compendex, INSPEC and others. In the area of grey literature, the DOE Information Bridge indexes grey literature produced by the DOE. Both full text and bibliographic records are available. The PrePRINT Network allows searches to be made across more than twenty preprint servers. Although most of these servers specialize in mathematics, computer science, and physics rather than in engineering, they do contain relevant resources for engineers.

In addition to the above resources, the DOE has also instituted a GrayLit Network. To quote, this site "makes the gray literature of U.S. Federal Agencies easily accessible over the Internet"[5] and the site is "the world's most comprehensive portal to Federal gray literature."[6] Although this is a worthwhile goal, serious researchers would do well to also utilize the search engines of individual federal agencies rather than relying solely upon the GrayLit Network. The site may be the most comprehensive portal for federal literature, but it does *not* appear to be the most comprehensive way to search the resources of an individual agency.

Miscellaneous U.S. Government Technical Reports

In addition to the agencies listed above, numerous technical reports are generated from other federal government agencies such as the Department of Transportation and the Nuclear Regulatory Commission. Some of these reports are distributed under the auspices of the GPO while others are handled by NTIS. Search the web sites of the specific agencies in order to determine if documents are available online, or use the GPO or NTIS databases to find the needed information to retrieve documents from other sources.

Some government-generated reports are also indexed in subject based databases. For instance, reports from the U.S. Bureau of Mines are indexed in GeoRef. The inclusion of government reports in subject databases is not consistent among databases, but they are a possible source of information.

State and Local Technical Reports

State and local agencies generate their own technical reports concerning matters of local interest. It may be a report on traffic flow for a particular stretch of highway, or a record of stream water temperatures at a generating plant site. In the case of these documents the method of retrieval can vary greatly depending upon the agency involved. Indexing on a national scale is often nonexistent, library depository programs vary or are nonexistent, and knowledge of these reports may be limited. While NTIS may include some of these, often the only way to retrieve the report is through direct contact with the issuing state or local agency.

TECHNICAL REPORTS–NON-GOVERNMENT REPORTS

Technical and research reports are also produced by non-government entities such as IBM and Bell Labs. These reports are frequently cited by corporate researchers who publish in commercially available research literature. Thus, awareness of the corporate reports, which normally have very limited distribution, is brought into the public arena. In some cases, providing the reports may not be difficult, because they are indexed in WorldCat and readily available through interlibrary loan. Other documents are not so easily located, and may require direct contact with the corporate publisher.

Some reports are easily located, but not so easily accessed. For instance, information concerning EPRI Technical Report # TR-106294, *An Assessment of Distribution System Power Quality* is easily located at the EPRI web site (http://www.epri.com). However, the 3-volume set is priced at $25,000 per volume. Thus, even though the report is identified and located, it may not be financially possible to access the contents.

STANDARDS

The International Organization for Standardization (ISO) states that:

> Standards are documented agreements containing technical specifications or other precise criteria to be used consistently as rules, guidelines, or definitions of characteristics, to ensure that materials, products, processes and services are fit for their purpose.[7]

Global manufacturing and communication has increased the need for worldwide standards. Products manufactured in one country must be operational in others. The format of the credit cards, phone cards, and "smart" cards that have become commonplace is derived from an ISO standard. The standard, which defines such features as an optimal thickness (0.76 mm), enables the cards to be used worldwide.

Standards can also play a role in litigation. It is not unusual for standards twenty or more years old to be requested as evidence in court cases. Reference to these superseded standards can show whether a manufactured item referred to in a lawsuit was produced in compliance with the standards in effect at the time it was manufactured.

The number of standards in use is large and increasing every year. In 1999 the number of American National Standards (ANS) increased by nearly 5.5% to a new total of 14,650 approved ANS.[8] At the end of 1999 the ISO had 12,524 standards in effect.[9]

Standards–Providing Access

In the past it has been difficult to provide standards access to researchers and practicing engineers. Onsite paper format collections were expensive to subscribe to, time consuming to update, and almost impossible to catalog efficiently. CD-ROM based full text collections raised the expense even higher, and often offered only current standards information, with no superseded standards available. The increasing availability of standards on the WWW has alleviated some of these problems and provides for greater flexibility.

Institutions that want unlimited access at a fixed cost can access standards through a flat rate annual subscription. This provides the greatest convenience for the end-user because the standards are available 24 × 7 to all affiliates of the institution. No intermediary is needed. For high use institutions where the cost of individual standards would exceed the flat rate subscription, or for research corporations where immediate access is a priority, the annual subscription plan is the preferred option.

Also available via the WWW is on-demand access to standards. For organizations that need only a few standards from any one issuing organization (e.g.,

ISO, IEEE), it is more economical to pay for the individual standards as needed rather than pay a flat rate for unlimited use. In most cases the "on-demand" system will require that standards orders be placed through a central office. Although central ordering will decrease flexibility, it will provide several benefits:

- Costs can be controlled and ordering abuses eliminated. Standards can cost hundreds or thousands of dollars each and uncontrolled purchasing could quickly deplete a deposit account.
- Records can be maintained of standards ordered. This can be particularly helpful if the standards are not cataloged for the collection but are given to the researcher. If another researcher later orders the same standard, referral can be made to the copy of the standard already being used by the previous requestor.
- If annual on-demand costs for a specific society routinely exceed annual subscription costs a subscription for the particular society can be initiated.

Following is a partial list of standards vendors:

Techstreet	http://www.techstreet.com/
Global Engineering Documents	http://global.ihs.com/
ILI	http://www.ili-info.com/us/
NSSN	http://www.nssn.org/index.html

All of these companies are worldwide in scope and are increasing the number of organizations from which they can offer on-demand online access. Some of these companies may offer features such as a standards update service, which notifies institutions whenever specific standards or standards meeting a profile are updated.

Verifying references to standards can be done using either online or paper indexes. The vendors listed above have online indexes that can be used to verify a standard's number, title, and publication year. In addition, organizations such as ISO and IEEE have online standards information at their web sites. The best paper index to standards is the *Index and Directory of Industry Standards*.[10] This seven-volume work contains both United States and international standards.

Providing standards by buying directly from the producing organizations or borrowing through interlibrary loan (ILL) may yield only limited results. Buying standards directly will only work part of the time because many producers only sell through such vendors as Global and ILI. Interlibrary loan (ILL) is often ineffective because libraries that have paper format collections are hesitant to lend standards. Also, as more libraries abandon the paper format in favor of online on-demand access, the pool of paper standards available for ILL will decrease.

Standards-Providing the Right Standard at the Right Price

As mentioned above, the costs of on-demand versus subscription based standards access should be monitored in order to assure the most cost efficient mode. For institutions buying on demand, there are three facts that should be noted:

- Each standard is unique. There is no "close enough" or "almost." If a spec calls for a particular standard, only that standard will suffice.
- Although each standard is unique, standards are often adopted and renamed among agencies. For instance, the ISO 14000 series of standards has been duplicated by many agencies. If a patron needs ISO 14001, then ASQ 14001 will also probably work. If standard ASQ 14001 is already in the collection, or if ASQ 14001 is less expensive than ISO 14001, the library can save money by supplying the equivalent standard. Some indexes note equivalent standards.
- Organizations often do not sell their own standards, but rely upon other companies to do this. Essentially, the organizations are the wholesalers and Global, ILI, etc., are the retailers. As with all retailers, there are differences in the mark-ups. It may pay to shop around.

MILITARY SPECIFICATIONS

Often referred to simply as "milspecs," this collection of documents dictates the specifications for items purchased by the United States Department of Defense (DoD). There are significant changes taking place with regard to these documents. During the past decade the military realized that its specifications often paralleled those that were being developed in industry. When military regulations were the same as industry standards, the military deactivated its regulations and referred contractors to industry standards. Along with the recognition that industry standards were legitimate substitutes for milspecs has come the recognition that products produced for commercial applications can be used in the military. The infamous $640 toilet seat and $435 hammer are not necessary and neither are the specifications responsible for their creation. The Defense Standardization Program of the DoD is placing much information about these changes online (http://www.dsp.dla.mil/). This site includes guidelines for using existing milspecs, links to full text milspecs online, and references to industry standards.

As the DoD relies more heavily upon industry standards, accepts commercial products more readily, and places an increasing number of its remaining specifications online there is a resulting library collection issue: the value of a full text subscription to milspecs from a commercial supplier such as Global Engineering is decreasing. In fact, subscribers to the online Standards Infobase from ILI now receive free online access to full text milspecs.

The reasons for these changes are clear. As the number of milspecs replaced by industry standards increases the number of full text "hits" available from a milspec subscription decreases. In the past a search in milspecs for a bearing specification would produce the full text milspec. The same search now may only produce a milspec cancellation notice and a reference to a Society of Automotive Engineers (SAE) standard. Unless the searcher also has a subscription to the SAE full text standard, the needed document is unavailable online. A milspec subscription only provides access to milspec full text, not to documents copyrighted by other organizations that are approved by the DoD. As the DoD adopts more industry standards, a milspec full text subscription will yield more citations for industry standards and fewer hits on full text milspecs.

The replacement of milspecs with industry standards also means that fewer standards used by the DoD will be freely available on the WWW. Whereas many milspecs issued by the DoD were freely available on the WWW, standards issued by other organizations and adopted by the DoD are only available through purchase. Thus, the amount of free access to standards utilized by the DoD is decreasing.

MEETINGS PAPERS/PREPRINTS/E-PRINTS

The above three terms have been grouped together, but there are slight differences between them. Meetings papers, as the phrase implies, are papers that are presented at a conference or other gathering. The papers may be published simply as a series of meetings papers, or may be published more formally in a conference proceedings or journal. The publication of meetings papers may take place at the same time as the conference, or if published in a journal may be delayed by as much as two to three years.

Preprints are also papers that are to be presented at a conference, but in the fields of computer science and physics the preprints are often distributed in hardcopy or electronically (e-prints) for discussion before the conference. Authors receive comments on the preprints and may make revisions based on those comments. Preprints may be published later in conference proceedings or as journal articles.

Preprints and e-prints are not as common in engineering as they are in the fields of computer science and physics. However, meetings papers are used by many professional organizations such as the SAE, American Society of Mechanical Engineers, and American Institute of Aeronautics and Astronautics. It is not possible to give detailed information on specific meetings papers series, but there are several peculiarities that should be kept in mind when considering them:

- Not all meetings papers are presented at a conference. It's possible that a paper may be written, but circumstances may prevent its presentation.
- Not all presentations that are given at a conference are available as a meetings paper. All too frequently a presenter does not submit a paper even though the presentation was made. Thus, just because someone heard a presentation given at conference is no guarantee that any written record of the presentation exists.
- Meetings papers are sometimes published in multiple versions. It is possible that a meetings paper will be published in the meetings papers series (technical papers series), in a related conference proceedings, and also as a journal article. For the collection developer, this means that care must be taken not to spend money on duplicate copies of the same papers. For the reference librarian, this means that it's possible for a paper to be cited in two or three different ways. Just because the library does not own the paper in the same form in which a publication cites it does not mean the library does not own the paper. It may be necessary to check multiple publications where the paper may be found.
- Distribution of meetings papers is problematic. Even papers that are published as a proceedings may not be distributed widely outside the conference attendees. In some cases complete papers are not published, but only titles and abstracts. This is particularly true of workshop sessions or informal presentations.
- Citations to meetings papers may be difficult to decipher for several reasons. Acronyms are often used and incomplete citations are common. Conference sponsors may change from one year to the next and sponsorship is often by multiple groups. Conference titles change frequently, and often the official bibliographic title as used in cataloging does not match the title commonly used or cited. Cataloging practices for proceedings also vary widely.
- Conferences vary in frequency. Some conferences may occur every year, while others are held on a biennial or even triennial schedule. Occasionally conferences are held only once, never to be repeated.

MANUFACTURERS' CATALOGS,
PRODUCT SPECIFICATIONS, DATASHEETS

A primary purpose of this literature is to inform consumers about the manufacturer's products with the hope that a sale can be made. Although it is seldom cited in literature, for a construction manager needing specs about the carrying capacity of a truck, or a computer engineer needing information on the pin configuration of a component, the literature is valuable. Librarians should be aware of its existence and availability.

In the past, libraries that have wanted to provide this material have had three choices: (1) freely available collections of second-hand and outdated volumes that were very difficult to maintain; (2) new, but still hard to maintain volumes purchased directly from manufacturers; or (3) commercially available collections on microfiche or CD-ROM that were easier to maintain but carried a hefty price tag.

Now however, many manufacturers' catalogs are available on the WWW, leading to some very desirable effects. Libraries that are relying on commercial providers for these materials are getting more product for their dollar. As more material is freely available on the WWW, commercial providers realize that the value of their material is decreasing and they need to lower subscription costs and/or provide more material and features. Libraries that have been depending upon non-commercial paper collections realize the end is in sight for those frustrating entities. Given that paper collections are often incomplete and out-of-date, the material on the WWW is now more than equal to the home-grown, hard copy data in many collections.

It appears that the trend for online access to manufacturer's information will continue. For libraries that cannot afford a commercial product, this means that the time is fast approaching when an investment of time will provide access to very good freely available material. For corporations that put a premium upon fast retrieval of information, the commercial providers will supply a well-indexed product with a standard interface for all manufacturers. Commercial providers may also be able to provide specialized indexing that will allow searches on specific parameters and comparison between products from competing suppliers.

Commercial Suppliers

As mentioned above, the simplest, but most expensive way to provide catalog data is through a commercial supplier. Two suppliers providing full images of manufacturers' catalogs through WWW accessible databases that allow searching by product or company name are:

IHS   http://www.ihs.com/
SoluSource   http://www.solusource.com/

Free Compilations

Some WWW sites offer free access to manufacturer's sites and catalogs. Thomas Register is a general database providing access to WWW sites in many different industries. Sweet's is a site that compiles construction-related suppliers. The detail and quality of the information provided varies greatly depending upon what the manufacturer places on the WWW.

Thomas Register
of American Manufacturers   http://www.thomasregister.com/
Sweet's   http://www.sweets.com/

Individual Manufacturer's Catalogs on the WWW

For researchers who know the manufacturer for the desired product the simplest tactic is often to go directly to the company WWW site. Many manufacturers are making their catalogs available on the WWW. As an example of the detail available, construction equipment manufacturers such as Deere and CaseIH have extensive web sites showing specifications on many of the products they manufacture.

Deere   http://www.deere.com/deerecom/default.htm
CaseIH   http://www.caseih.com/construction

Component manufacturers such as Intel and Micron offer extensive online datasheets for their products.

Intel   http://www.intel.com
Micron   http://www.micron.com

PATENTS AND TRADEMARKS

U.S. Patents and Trademarks

United States patents have become readily available on the WWW. The IBM site, now offered by Delphion, was the first provider of full patent images on the WWW, and offered a backfile to 1971. Recently, the U.S. Patent and Trademark Office has posted full page images of all U.S. patents on its web site (http://www.uspto.gov/patft/).

Trademarks from the U.S. are also readily available. The U.S. Trademark Electronic Search System (http://www.uspto.gov/web/menu/tm.html) offers several search options and provides links to online images of the trademarks.

Patents and Trademarks from Other Countries

Patents and trademarks from countries other than the U.S. are not quite as easily obtained. The Delphion site (http://www.delphion.com/), offers access to European patents and patent applications, as well as patent abstracts of Japan.

Searching for Patents and Trademarks

In one major respect patents and trademarks are unlike most other literature. With most literature, a search of subject databases, bibliographies, or other tools will provide an extensive and rather complete set of citations. The major problem is obtaining the cited document.

With patents and trademarks, it is just the opposite. The major problem is searching the literature, navigating the various classifications, and making sure that all previous documents related to the new invention or trademark have been ferreted out. These search techniques take extensive training and much experience to master. Once the list of relevant documents has been formulated, online access makes retrieving the documents simple.

MATERIAL SAFETY DATA SHEETS

There are numerous products in the construction and manufacturing industries that can be hazardous to one's health. Solvents, pressure treated lumber, particle board, paints, and many other substances can be dangerous if not used properly. Material safety data sheets (MSDS) can be a valuable source of information regarding the proper work procedures for these products.

The Canadian Centre for Occupational Health and Safety (CCOHS) defines a material safety data sheet (MSDS) as follows:

> A Material Safety Data Sheet (MSDS) is a document that contains information on the potential health effects of exposure and how to work safely with the chemical product. It is an essential starting point for the development of a complete health and safety program. It contains hazard evaluations on the use, storage, handling and emergency procedures all related to that material. The MSDS contains much more information about the material than the label and it is prepared by the supplier. It is intended to tell what the hazards of the product are, how to use the product

safely, what to expect if the recommendations are not followed, what to do if accidents occur, how to recognize symptoms of overexposure, and what to do if such incidents occur.[11]

Until recently it could be very difficult to obtain MSDS. It was usually necessary to contact the manufacturer of the product and delays were common. Now, however, many are available on the WWW.

A search on Alta Vista or a similar search engine using the strategy:

+"material safety data sheet" +"company or material name or brand name"

will often yield results. For example, a search statement such as:

+msds +Sheetrock or +"material safety data sheet" +"pressure treated lumber"

will yield good results.

A search within subject directories such as Yahoo or Lycos can also be profitable.

THESES AND DISSERTATIONS

For many years, dissertations have been available through two main sources: interlibrary loan (ILL) from the granting institution or photocopies from the UMI service. Master's theses, which are not as comprehensively handled by UMI, were ordinarily available only through ILL from the granting institutions.

During the past few years, some institutions have made full text theses and dissertations (electronic theses and dissertations-ETDs) available on the WWW. This format change has enabled ETDs to be accessed instantaneously. It has also raised questions about copyright and the ability of ETDs "published" on the WWW to be published later as journal articles, books, or technical papers. This has caused some authors of ETDs to restrict the distribution of their ETDs on the WWW. The final resolution of this problem is still unknown but the track record of preprints in computer science and physics shows that WWW published materials still have value as peer-reviewed journal articles published later. Hopefully this will be true with ETDs as well.

INTERNATIONAL RESOURCES FOR GREY LITERATURE

While the preceding sections have dealt primarily with United States resources, grey literature flourishes worldwide. There are several information resources available that may be helpful in providing information on the worldwide scope of grey literature.

The *International Journal on Grey Literature*, published by MCB University Press provides general information about grey literature as well as articles oriented to the sci/tech community.

The SIGLE (System of Information for Grey Literature in Europe) database provides information on a wide range of grey literature formats including research reports, preprints, committee reports, working papers, dissertations, and conference papers. The SIGLE database is managed by the European Association for Grey Literature Exploitation (EAGLE), a fifteen-country organization formed to improve access to European grey literature (http://www.kb.nl/infolev/eagle/frames.htm).

The GreyNet site (http://www.greynet.org/), connected with MCB University Press, offers information about the Grey Literature conferences as well as general information about grey literature.

CONCLUSION

Several types of engineering grey literature have been discussed, and specific examples of grey literature have been described. However, the rapid changes in information technology guarantee that access methods and availability of grey literature will change quickly.

The engineering librarian will find it necessary to constantly reevaluate the methodology for obtaining this material. Sources such as the *International Journal of Grey Literature* will be helpful, but publishing lag times mean that the latest access changes will be apparent only to those who seek them out. Commercial database providers and journal publishers vie for subscription dollars and eagerly send out their representatives to tout their latest developments. The publishers of grey literature are generally not so aggressive in their marketing. Indexes are developed, access is improved, and documents are placed on the web, but in many cases only the librarians that proactively seek out the literature will be aware of these enhancements.

REFERENCES

1. GreyNet, [cited November, 2000]. http://www.greynet.org/about/greylit.html.
2. C. P. Auger, *Information Sources in Grey Literature*, 4th ed. (London: Bowker Saur, 1998), 3.
3. Ibid.
4. Ibid.
5. GreyLit Network [cited November, 2000]. http://www.osti.gov/graylit.

6. Ibid.

7. ISO [cited November, 2000]. http://www.iso.ch/infoe/intro.htm#What are standards.

8. American National Standards Institute [cited November, 2000]. http://web.ansi.org/public/about.html.

9. ISO [cited November, 2000]. http://www.iso.ch/infoe/iso_in_figures.pdf.

10. Information Handling Services, *Index and Directory of Industry Standards* (Denver, Colorado: Global Professional Publications).

11. CCOHS [cited November, 2000]. http://www.ccohs.ca/oshanswers/legisl/msdss.html.

International Resources
in Science and Technology:
A Review with Two Case Studies

Bonnie A. Osif

SUMMARY. Access to international resources is important to researchers in science and technology. Often these resources are government reports, publications from international agencies and associations and conferences. Many of these resources are difficult to obtain or borrow in an efficient manner. While various organizations have attempted to collect in problem areas, a number of subjects and format types are still problematic. The need for international resources is reviewed and several international collection projects are described. Agriculture and transportation provide two case studies to indicate the strengths and weaknesses of their collection strategies. *[Article copies available for a fee from The Haworth Document Delivery Service: 1-800-342-9678. E-mail address: <getinfo@ haworthpressinc.com> Website: <http://www.HaworthPress.com> © 2001 by The Haworth Press, Inc. All rights reserved.]*

KEYWORDS. International literature, virtual libraries, National Transportation Library, National Agricultural Library

It is difficult to pick up a newspaper, magazine or journal that does not have some information on globalization. Whether the specific topic is transnational

---

Bonnie A. Osif, BS, MS, EdD, is Engineering Reference Librarian, Penn State University, and Librarian, Pennsylvania Transportation Institute.

Address correspondence to the author at: Penn State University, 325 Hammond Building, University Park, PA 16802 (E-mail: bao@psulias.psu.edu).

[Haworth co-indexing entry note]: "International Resources in Science and Technology: A Review with Two Case Studies." Osif, Bonnie A. Co-published simultaneously in *Science & Technology Libraries* (The Haworth Information Press, an imprint of The Haworth Press, Inc.) Vol. 19, No. 3/4, 2001, pp. 75-86; and: *Engineering Libraries: Building Collections and Delivering Services* (ed: Thomas W. Conkling, and Linda R. Musser) The Haworth Information Press, an imprint of The Haworth Press, Inc., 2001, pp. 75-86. Single or multiple copies of this article are available for a fee from The Haworth Document Delivery Service [1-800-342-9678, 9:00 a.m. - 5:00 p.m. (EST). E-mail address: getinfo@haworthpressinc.com].

corporations, international pollution, international collaboration or the European Union, it is a topic that demands our attention.

As science and technology librarians, there are several aspects that affect us directly. These include increased international collaboration by our patrons, increase in the number and breadth of databases with the resultant knowledge of an increased pool of resources, and the growing realization that best practices and technological advances don't recognize political boundaries. While the awareness of international venues, cooperative ventures and resources has increased, access has not kept pace. In the middle is the librarian, bridging the gap between demand and access by developing innovative strategies.

This paper will address international information by noting studies on the importance of these resources, reviewing some quantitative data and then presenting two contrasting case studies with preliminary recommendations.

WHY INTERNATIONAL RESOURCES?

The above question has been voiced with the supporting statement that the U.S. is a key technological leader and any worthwhile research will be published in English and in a resource that is routinely available in the U.S., either in-house or through routine interlibrary loan.

There is actually some validity to this point of view. There is a wealth of literature concerning journal rankings and the importance of publishing in the major journals or presenting at the major conferences. There is strong international showing in these venues, therefore it logically follows that the best, most worthwhile international research is being published in resources that the U.S. researcher can readily obtain.

However, this argument is not as strong as some might believe if viewed from a less narrowly focused perspective. Granted many researchers, especially academic, aim for these prestigious venues to present their research. However, many researchers do not, for various reasons, attend, present or write for these. Often, a finding is relegated to a report with normal distribution to the sponsoring agency and a few other interested parties. It may not be indexed with the only access point a note in a list of references at the end of a paper. Many of the materials that fall into this category are grey literature, international, or both.

If collection rate is low and indexing spotty, how will patrons learn of the material they are missing? And, if they don't know about it, is it important enough to be of concern to the librarian?

Research is no longer geographically isolated. International collaboration is nothing new but the Internet has made it even easier, faster and cheaper. Cost is another factor in the need to know about research, regardless of geographic boundary. "Reinventing the wheel" is costly in time, money and effort. While the quote is discipline specific, the U. S. Secretary of Transportation succinctly makes the point by charging the transportation programs to "inform the United States highway community of technological innovations in foreign countries that could significantly improve highway transportation in this country."[1] The same charge could be given to any technological discipline.

Research needs often require rapid delivery. While the particular situation creates differing needs and there may be some variations based on subject matter, a number of articles have made reference to the information seeking patterns of engineers that have direct implications for international information.[2,3] Engineers tend to use their own knowledge or that of a colleague, resources already in their collection or in that of a colleague, or otherwise readily available. They often are not strong library users, neglecting to utilize the expertise of the engineering librarians who might be able to identify and obtain valuable information resources. Long lags in obtaining interlibrary loans, or trying to find alternative routes for materials not available through normal channels are not acceptable. Therefore, physical or electronic "ownership," or very efficient interlibrary loan or document delivery is a necessity. In addition, "proactive marketing" of library resources, research skills and the myriad information resources that are available has become crucial.

THE VALUE OF INFORMATION

There is significant research on information and the value of information services. While many are anecdotal, some provide mathematical formulas and numerical analysis. Results from these studies vary and readers will need to interpret and evaluate the different perspectives. However, the comparison between value and cost is informative. A number of articles have provided a summary on the topic of the value of information for those interested in further investigation.[4,5]

The Special Libraries Association (SLA) has devoted considerable attention to the valuation of services since many corporate and special libraries are required to justify their existence to their administrations. *Valuing Special Libraries* [6] is one such study. Griffiths and King[7] provide values for return of investment ranging from 7.8:1, to 14.2:1. *Competencies for Special Librarians of the Twenty-First Century*[8] submitted to the SLA Board is an up to date look at the value of information specialists.

Dresley and Lacombe (1998)[9] did a study for the U.S. Department of Transportation, Washington, D.C., in the discipline of transportation. It is a study that provides a template for others to follow. There are a number of very specific scenarios with dollar amounts attributed to the use of quality information generated from the library. Two of the examples include the saving of $300,000 in strengthening of steel bridges, and nine million saved in annual life cycle costs of bridge deck concrete. This last case stated that the savings came from "mining" best practices and ideas from literature reviews.

The generalization derived from these varied studies is that the value of mining quality timely information far outweighs the costs. Efficient access to pertinent information can save money, time, effort and increase safety, productivity and competitiveness.

FORMAT CHARACTERISTICS OF SCIENCE AND TECHNOLOGY LITERATURE

Access to the literature of science and technology appears to be more problematic than many other disciplines. The reasons are myriad and vary somewhat by discipline. They include a significant number of publications by professional societies, limited advertising of some publications, lack of indexing, distribution of conferences only to on-site registered attendees, limited timeline for ordering a publication, and a preponderance of grey literature. At the same time there is a decrease in international exchanges and decrease in non-English language collections. In addition, long term stress on serial budgets, and a decrease in monograph budgets affect the ability to purchase lesser known conferences and peripheral monographs.

While each reader will have a wealth of individual examples of difficult, if not impossible, to obtain resources, statistical studies are more objective and convincing. These studies include evaluation of bibliographies for formats and availability.

One such study was done by Musser and Conkling.[10] While it only listed basic statistics and did not provide discipline specific numbers, it does provide a large sample set of 4,780 references which measured the literature format of 212 articles from sixteen engineering journals. According to this study, journals accounted for 53% of the references, conference papers 19%, monographs 12%, reports 9%, dissertations and theses 3% and other resources 4%.

Edwards[11] studied the references of thirty-two theses and dissertations in polymer sciences. The resulting 3,648 references produced 72.8% journals, 15.4% monographs, 4.6% patents, 3.3% proceedings, 1.9% dissertations, .5% standards and 1.5% other, which included reports.[12]

In preparation for this article a short "check" survey was done on 394 references, approximately 100 each from *IEE Proceedings Vision, Image and Signal Processing, Integrated Computer-Aided Engineering*, and the *Journal of the Air and Waste Management Association, Journal of Guidance, Control, and Dynamics*. The results were 44% of the references were to monographs or conference papers, 46% to journals, 7% to reports, 2% to dissertations/theses and .5% each to web sites and personal communications. These references were checked in OCLC for the number of holding locations. The study indicated that only 2% of the books were not listed, whereas 85% of the reports were not listed in OCLC or NTIS.

There is variation in the numbers cited in the above studies, some of which might be accountable by the subject. However, they all indicate the wide range of resource types used. There is, however, further study that might be interesting and informative: how does the bibliography relate to records indexed in the appropriate databases and to reasonable accessibility of the actual items.

In comparison to the above studies of bibliography, a series of studies investigated formats cited and indexed in transportation. Two were based on database records generated by subject searches in TRIS (Transportation Research Information Service) and Transport (TRIS, plus Transdoc and IRTD databases). Although the results varied somewhat, the importance of report literature was high. Two other transportation studies were similar to the Edwards and Musser/ Conkling studies of bibliographies. The first study used Transportation Research Board (TRB) publications, the second a random sample of European reports published in Finland, Germany, The Netherlands, Sweden, Switzerland and the United Kingdom. Table 1 summarizes the results of the four studies.

In addition to coding by format, citations were coded for language. Table 2 summarizes the English/non-English distribution in each study. Citations were checked in OCLC and NTIS for the number of holding locations in the U.S. and Canada. The results are summarized in Table 3.

Results were extremely discouraging. The four studies reinforce the belief that a significant amount of international, especially non-English language resources are not readily available for efficient access. Report literature in particular is difficult to obtain.

TABLE 1. Format Types

Format	TRIS Study (1995)	Transport Study (1996)	References– TRB Publications	References– International Publications
Journals	28.4%	22.2%	30.7%	12%
Books/Conferences	21.6%	41.7%	39.3%	26%
Reports	49.6%	35.7%	28.7%	58%
Theses	.4%	.4%	1.3%	1%
Others	0	0	0	3%

TABLE 2. Language Distribution of References

	TRIS Study (1995)	Transport Study (1996)	References– TRB Publications	References– International Publications
English	91%	83.3%	96%	36%
Non-English	9%	16.6%	4%	64%

TABLE 3. Holding Locations Noted in OCLC

	TRIS Study (1995)	Transport Study (1996)	References– International Publications
0	23.4%	23.3%	72%
1	3%	12.3%	7%
2+	73.6%	64.4%	21%

EXISTING PROJECTS

Patrons, value of information studies and increased globalization support the importance of international publications. A number of studies indicate the possible difficulties in accessing them. To address the need for international information there are a number of initiatives with the goal to insure improved access. Many are subject, country or format focused and a number are in non-science and technology areas. A few examples in the science and technology subjects will be described briefly to give some framework for the two case studies that follow.

The Science Research Materials Project at the Center for Research Libraries collects materials by publication source rather than subject. They collect over 500 Japanese serials, serials and monographs from the Library of the Russian Academy of Sciences, serials from South and Southeast Asia governments and dissertations from a number of foreign universities. The collection can be searched on their website [http://wwwcrl.uchicago.edu/info/srmp/].

The Canadian Institute of Science and Technology Information (CISTI) has one of the largest collections in North America with approximately 500 new items coming into the system each day. CISTI is developing strong collections or partnerships with Sunmedia of Japan for reports and articles, Korea Institute of Industry and Technology Information for Korean journal articles, Institute of Scientific and Technical Information of China, People's Republic of China for research journals, monographs, conference proceedings and reports, and

Science and Technology Information Center in Chinese Taipei for journal articles. Since few libraries can afford to subscribe to this range of Asian materials, it provides a very valuable service.

The Institute of Transportation Studies (ITS), University of California, Berkeley, one of the foremost transportation libraries in the world, has a number of very good exchange and collection programs with international organizations. These include: Bundesanstalt fur Strassenwesen (Federal Highways) Germany, Institut National de Recherche sur les Transports et leur Securite France, and the Finnish National Road Administration. Informal exchanges exist with ARRB Transport Research, Ltd. Australia and Quebec Ministere des Transports Canada.

Virtual libraries are an increasingly popular option. There are a number of examples, which include The Virtual Technical Report Center at the University of Maryland, a collection with "links to technical reports, preprints, reprints, dissertations, theses, and research reports of all kinds."[13] The Networked Computer Science Technical Reference Library (NCSTR) is a cooperative project of approximately one hundred institutions which links an "international collection of computer science research reports and papers."[14] Many of these reports would be difficult to obtain without the websites.

These are worthy attempts to improve access to international information resources. However, there are some major issues that are not addressed. The major focus of attention tends to be developed countries in Europe and Asia, major associations, and serials. Collection in African and South American resources is low, smaller associations may be ignored and grey literature is still collected and cataloged very inefficiently. Even collection patterns are "haphazard." Some libraries collect by country, some by topic, some by publisher and others by a combination of methods. Coordination is low. This allows a great deal of information to fall through the cracks as there is not an overseer to locate the gaps in the collection and designate a collector.

One other important consideration is the trend toward electronic access and virtual libraries. This trend is commendable, efficient and popular, but there are many items, including technical reports, that cannot be put on the Web free of charge. A number of international organizations must charge for their publications as they are self-supporting units. While this is contrary to the U.S. trend to put government documents on the web, it is a reality that has important implications to the research community.

TWO CASE STUDIES: AGRICULTURE AND TRANSPORTATION

To illustrate the current state of collection this paper will look at two different subjects and their approaches to solving the problem of access to the information resources. Both disciplines share a strong international emphasis and format diversity. Both have "national" libraries in the U.S. However, they have, at least in the past, followed different paths and a comparison will illustrate strengths, weaknesses and possible scenarios for the future access of information.

Agriculture is one of the largest segments of the economy. Information needs range from economics, agronomy, biology, entomology, animal sciences, pathology, food science, engineering, and genetics, with social, cultural and environmental issues part of the overall picture. There is a strong practical aspect to the field of agriculture and the resources are varied, ranging along a continuum from academic interest to practical application, in conference papers and journals, to government regulations, standards and extension service circulars, brochures and bulletins for the practitioner. Individuals will have widely divergent interests in varying segments of this continuum based on their particular expertise and project.

In addition to a wide range of topics that comprise the field of agriculture and the breadth and depth of literature formats needed for the practice of this science, there is a strong time constraint that pervades the discipline. Crop disease, insect infestation, weather stress on crops and other commonplace occurrences do not always lend themselves to careful, time-consuming research and lengthy interlibrary loan requests. However, rash, poorly researched action may be useless, counterproductive or dangerous. Therefore, quality validated information is needed in a timely fashion and it is somewhat difficult to predict exactly what will be needed.

Linked with the need for quality information is a growing acknowledgement that there is a strong global aspect to agriculture. Whether it is the genetics of original seed potatoes or natural resistance to a pest, valuable information comes from all corners of the world. Much of vital importance can be learned from areas that are not very advanced technologically, while modern methods have proven their value in developing countries.

In addition, agricultural trade is global and there are many issues that must be addressed. The British beef situation of several years ago and the current protest against genetically altered foods are two illustrations.

To address these complicated needs for a wide range of information from a number of locales, in many languages, different literature formats and a broad range from the high level research level to the very practical application, a complex and coordinated system is needed.

The National Agricultural Library (NAL) of the U.S. Department of Agriculture (USDA) is one of the four national libraries (medicine, education and the Library of Congress are the other three, and others are in the planning stage). It strives to acquire, organize, manage, preserve and provide access to information, maximize access and enhance global cooperation. One of the listed values is "we believe that universal access to information is critical to the continued development of agriculture."[15]

The NAL has a very active international focus and has developed a program to make the statement more than a vision, but a reality. They have developed a good international exchange program. There are sixteen centers around the world in the Consultative Group for International Agricultural Research (CGIAR), including Peru, the Philippines, India and Kenya, which send their publications to the NAL. In addition, NAL has created over 7,000 relationships with governments, organizations and institutes of higher education who send their publications to the NAL. This program brings over 70,000 items per year to the collection. Formats range from the traditional library resource to the more difficult to obtain reports literature, working papers, statistics and more.

The scope of the collection is broad. They collect comprehensively in crop and soil sciences, veterinary science, food and nutrition, agricultural economics and rural sociology and have strong holdings in botany, chemistry and entomology. In addition to these historically strong collections, the NAL is dedicated to addressing issues of importance when they arise such as global warming, biodiversity, invasive species, etc. One of the obvious strengths of the NAL's collection development policy is the wealth of grey literature that is collected.

Access to the materials in agriculture is through AGRICOLA, the premier database in the field. It indexes over three million items from 1979 to the present, including journal articles, books, book chapters, USDA reports, State Experiment Station publications, and State Extension service publications. The database is available on the web or through vendors. It should be noted that not all of the international materials are included in AGRICOLA.

Transportation accounts for 11% of the Gross Domestic Product and is the fourth largest component in that index.[16] Like agriculture, the necessary resources ranged in format type from conference papers, journal articles, books, standards, and specifications to technical reports, white papers, and working papers from a wide range of organizations, associations, consulting firms and governments, both local, national and international. A very large percent is grey literature. In the 1995 and 1996 database studies by Osif, 73% and 85% of the records were to grey literature.[17]

Like agriculture, transportation encompasses a broad range of subjects. These include areas of civil and mechanical engineering, material sciences, ergonomics, environmental sciences, psychology and business. Time is a constraint, as is a very real need for cost effectiveness and "best practices," the knowledge of what has been done well by others to save time and money. As in agriculture there can be a very real dichotomy between the researcher and the practitioner, with the latter requiring fast access to a practical summary and the former requiring thorough, theoretical information.

To address these needs the Transportation Equity Act for the 21st Century (TEA 21) legislation authorized the formation of a National Transportation Library (NTL). The library is a virtual library, with full text documents and databases available through a free web site. Searching is through a web search engine and the major transportation index, TRIS from the Transportation Research Board, is available through the web site. The NTL is the responsibility of the Bureau of Transportation Statistics, one of eleven individual operating administrations of the Department of Transportation (DOT). The documents available through the NTL are in the public domain. Most are documents from the various administrations of the DOT although documents from all other organizations that can be mounted without copyright restrictions are requested.

The goal of NTL is commendable, however adequate funding has been deficient from the inception and the mission has not been clearly articulated. At the writing of this article a new head of the NTL had begun to address these issues.

One of the issues that must be addressed, although possibly not solved, by the NTL is the overall role it can play in access to transportation literature. At this point there is no NAL equivalent with centers collecting materials and sending them to a common, designated location for access by the community of researchers. The collection of transportation information is scattered, driven more by local policies, interests and resources with only "volunteer" coordination. The driving force has been the Transportation Division of the Special Libraries Association. Unlike agriculture, there is no national coordinator or true clearinghouse.

The NTL, while an ambitious, modern and well-intentioned attempt to collect and organize the literature, has some issues that need to be addressed. First, it is a virtual library and a significant percent of the materials needed by the transportation community cannot be accessed through the site. Many resources are copyrighted, including many of the desperately needed international reports. These resources will still need to be collected in print format, indexed, and made available to the transportation community.

Second, the NTL needs funding and staff to fulfill its assigned tasks: finding, organizing, linking and maintaining the electronic documents that are eligible for inclusion. While this sounds routine, the studies of transportation literature indicated that a significant amount of literature generated by local and state transportation organizations in the U.S. still are difficult to obtain. This must be systematically addressed if the trend is not to continue with the electronic library. Third, the archiving of documents is essential.

Last, the task of a coordinated effort to guarantee efficient access to international resources must be pursued. There are models from the NAL, the National Library of Medicine and others. These can serve as a guide for a non-virtual, print oriented segment of the NTL that is needed to address the total transportation information needs of the country. There is a project currently funded by the National Cooperative Highway Project, "Accessibility of Non-English Language Transportation Information," to evaluate options in both the collection and translation of non-English materials.

The need is evident, the solution complex. The final scope and format of the NTL is still not known. The hope is strong that the issues can be addressed and the transportation field will have access to the wealth of information available beyond U.S. borders. To accomplish this goal ongoing adequate funding and staffing, collection development, indexing, coordination and a patron friendly website is essential. A combination of electronic access, print collections and fast and efficient document delivery must be made available. The NTL may provide a lesson for other national libraries that are in some stage of creation or development.

CONCLUSION

Gregor Mendel published the results of his famous genetic crosses in 1866 in the *Transactions* of the Natural Science Society of Brunn. It was not until 1900 that the findings were rediscovered and biology was revolutionized.[18]

The advent of the web has provided the means to provide at least a partial solution to the problem of identifying resources and then obtaining them. Web databases and full text resources can provide relatively easy means of access. Models such as the NAL and the developing NTL can be expanded to other subject areas and to a wider range of resources. This is not a new idea. In 1981 Breton wrote "Since defective and noncompetitive technology is the outgrowth of deficient knowledge, it would appear that the systematic development and fundamental reliance on a knowledge base to support the research, development, and engineering efforts would assume priority over all other management considerations."[19] That knowledge base is now a major focus of attention with considerable effort to improve information access.

Science and technology cannot afford to hide treasures of discovery and innovation. Access, systematic and thorough, is necessary to the scientific and technological advancement. The electronic environment has given us some means to address these needs. However, planning, funding and cooperation are still necessary for effective access.

REFERENCES

1. TEA-21-Transportation Equity Act for the 21st Century in TEA-21 Moving Americans into the 21st Century [electronic document] (Washington, D.C.: FHWA, 2000)[cited October 25, 2000], available at *http://www.fhwa.dot.gov/tea21/h2400-v.htm*.

2. Holland, Maurita Peterson and Christina Kelleher Powell. "A Longitudinal Survey of the Information Seeking and Use Habits of Some Engineers." *College and Research Libraries* 56 (1995): 7-15.

3. Pinelli, Thomas E. "The Information Seeking Habits and Practices of Engineers." *Science & Technology Libraries* 11 (1991): 5-25.

4. Keyes, Alison M. "The Value of the Special Library: Review and Analysis." *Special Libraries* 86 (1995): 172-187.

5. Osif, Bonnie A. and Richard L. Harwood. "The Value of Information and the Value of Librarianship." *Library Administration and Management* 14 (2000): 172-178.

6. Kantor, Paul B. and others. *Valuing Special Libraries* Washington, D.C.: Special Libraries Association, 1998.

7. Griffiths, Jose-Marie and Donald W. King. *Special Libraries: Increasing the Information Edge.* Washington, D.C.: Special Libraries Association, 1993.

8. "Competencies for Special Librarians of the 21st Century" [electronic document] (Washington, D.C.: Special Libraries Association, 1996), [cited October 4, 2000], available at *http://www.sla.org/content/professional/meaning/competency.cfm*].

9. Dresley, S.C. and A. Lacombe. *Value of Information and Information Services.* Washington, D.C: U.S. Department of Transportation, 1998.

10. Musser, Linda and Thomas W. Conkling. "Characteristics of Engineering Citations." *Science & Technology Libraries* 15 (1996): 41-49.

11. Edwards, Sherri. "Citation Analysis as a Collection Development Tool: A Bibliometric Study of Polymer Science Theses and Dissertations." *Serials Review* 25 (1999): 11-20.

12. Edwards, Sherri. Electronic letter to the author. 4 October 2000.

13. The Virtual Technical Report Center [electronic database] (College Park, MD: University of Maryland, 1999) [cited October 4, 2000], available at *www.lib.umd.edu/ UMCP/ENGIN.TechReports/Virtual-TechReports.html*.

14. The Networked Computer Science Technical Reference Library [electronic database] (cited October 4, 2000), available at *cs-tr.cs.cornell.edu*.

15. Dubois, Rae. Electronic letter to author. 6 October 2000.

16. *Transportation Statistics Annual Report 1999.* Washington, D.C.: Bureau of Transportation Statistics, 2000 [Cited October 4, 2000], available at *http://www.bts. gov/transtu/tsar/tsar1999/summary.htm*.

17. Osif, Bonnie A. "Transportation Information: A Review of Grey Literature by Format, Language and Availability." *International Journal of Grey Literature* 1 (2000): 12-17.

18. "Mendel, Gregor (Johann)" In *Encyclopedia Britannica Online* [encyclopedia online] [Cited October 4, 2000], available at *www.eb.com:180/bol/topic?eu=53267& sctn=1*.

19. Breton, Ernest J. "Reinventing the Wheel: The Failure to Utilize Existing Technology." *Mechanical Engineering* 103 (1981): 54-57.

Engineering Resources for Children–Kindergarten Through 12th Grade: A Case for Dispositional Learning

Justina O. Osa

Steven L. Herb

SUMMARY. This article utilizes a long-established model for early learning to examine school-age behaviors associated with future engineering skills. Originally developed by Lilian Katz, the four learning components of the model include knowledge, skills, dispositions, and feelings. Interactive learning is described as the most successful form of instruction. It appears that dispositional learning or learning habits are a better focus for kindergarten through 12th grade engineering instruction, rather than specific knowledge about engineering or specific engineering skills. Engineering programs in elementary and secondary schools, web resources, and children's books are sampled and described in four behavior categories associated with engineering–Thinking in New Ways;

Justina O. Osa, BA, MEd, MSLS, EdD, is Education and Behavioral Sciences Librarian, Penn State University Libraries, University Park, PA 16802. Steven L. Herb, BS, MEd, MSLS, PhD, is Head, Education and Behavioral Sciences Library and Affiliate Associate Professor of Language and Literacy Education, Penn State University, University Park, PA.

The authors wish to thank Elaine Hannon for research assistance with this article.

[Haworth co-indexing entry note]: "Engineering Resources for Children–Kindergarten Through 12th Grade: A Case for Dispositional Learning." Osa, Justina O., and Steven L. Herb. Co-published simultaneously in *Science & Technology Libraries* (The Haworth Information Press, an imprint of The Haworth Press, Inc.) Vol. 19, No. 3/4, 2001. pp. 87-103: and: *Engineering Libraries: Building Collections and Delivering Services* (ed: Thomas W. Conkling, and Linda R. Musser) The Haworth Information Press, an imprint of The Haworth Press, Inc.. 2001. pp. 87-103. Single or multiple copies of this article are available for a fee from The Haworth Document Delivery Service [1-800-342-9678, 9:00 a.m. - 5:00 p.m. (EST). E-mail address: getinfo@haworthpressinc.com].

87

Finding Out How Things Work; Designing and Building and Disassembling and Rebuilding; Dreaming and Imagining. *[Article copies available for a fee from The Haworth Document Delivery Service: 1-800-342-9678. E-mail address: <getinfo@haworthpressinc.com> Website: <http://www.HaworthPress.com> © 2001 by The Haworth Press, Inc. All rights reserved.]*

KEYWORDS. Dispositional learning, engineering resources, K-12 resources, instructional activities, curriculum resources, active learning

Scientists discover the world that exists; engineers create the world that never was.

–Theodore von Kármán

INTRODUCTION

Engineering is the art or science of making a practical application of knowledge, principles, experience, and judgment; of pure sciences and common sense. Engineers, or the best engineers, must be creative and be able to direct their creativity toward solving problems. They must be able to analyze and dream; synthesize and imagine; stop and find new ways to start; create new things that solve old problems; and make old things work in new ways.

We are in a period of unprecedented and rapid social and technological changes, and it is imperative that engineers think more creatively and more quickly than ever before. It is only through ingenious thoughts and the inventions those thoughts produce that the world of the 21st century can cope with and adapt to the pace of change. Society expects that students will acquire the knowledge, skills, and attitudes that are necessary for our culture to continue to gain deeper insights into the phenomena that make up our universe, and to solve "real-world" problems. Therefore, teachers must aim higher than merely transferring particular bits of knowledge, situation specific skills, solid work habits, or good attitudes toward learning to their students–teachers must find ways to synthesize all four types of learning into a constantly evolving young person. Teachers must assist students in their intellectual growth, competency development, and self-discovery. Teachers have the great and challenging task of preparing students to deal with the world of today while also preparing them for the world of tomorrow. The complex question is, how are students to be prepared to become creative thinkers and ingenious problem solvers?

Teachers must stop doing business as usual and explore new methods that will produce the types of learning, indeed, the types of students that the world is demanding. One particular method of conducting classroom business in a new way grows from the discussion of constructivism in education. Constructivism is a theory, based on research from cognitive psychology, that advocates that people learn by constructing their own knowledge through an active learning process, rather than by simply absorbing knowledge directly from another source (Ryan and Cooper, 1998).

A research-based knowledge of how students learn would enhance the success of the teacher's search for appropriate methods of instruction that will produce the critical, creative thinker and problem solving individual society wants schools to produce. Teachers need to know how students create, or construct their own knowledge, and the conditions that impact their construction of that knowledge. The passive acceptance of information leading to knowledge is passé. Knowledge must be actively constructed by students–and based upon a personal context of prior knowledge, skills, dispositions, and feelings (Katz, 1999). The learner must construct meaning from experience. Students identify and construct guidelines by experimenting, examining models, reflecting, and deciding on functional patterns that fulfill their personal needs. The methods teachers espouse to successful effect within this theory should be those that enable them to become guides rather than directors.

LEARNING MODEL FOR SCHOOL AGE INSTRUCTION IN ENGINEERING

When looking at instructional resources for school age children it is nearly always helpful to categorize the types of learning those resources will support. Although this article focuses on engineering, we are actually attempting to identify those school-age behaviors that might be a good predictor of the successful acquisition of engineering knowledge, skills, and behaviors at the college level. A look at a learning model developed by the early learning specialist Lilian Katz may prove helpful in this categorization of types of learning, and in how we examine the programs and resources that are the focus of this article. Lilian Katz has been active in the field of early learning for several decades, but in recent years her work has been applied to emergent print literacy in school age children and information literacy in college age students. The three key aspects of Katz's learning model that make it applicable to nearly any emerging learning are:

- A focus on what is learned when–the importance of what children can learn vs. what they should learn;
- A focus on the learning of knowledge, skills, dispositions and feelings; and
- A focus on collaboration. (Katz, 1999).

A closer look at these three aspects of Katz's model will help frame our examination of engineering programming and resources in and for elementary and secondary schools.

What Is Learned When–Capability of Children vs. Suitability of Instructional Timing

Many years ago, one of the authors was observing children's librarians in a public library conducting library tours that focused on how to find information. One after the other demonstrated to fourth grade children how one found books about rocks or buildings or novels by Judy Blume, followed by a bit of catalog exploration by the kids, followed by a five minute story. Some of the kids were interested in Judy Blume or rocks or building, but many more of the kids were interested in getting a drink of water or wondering if they could photocopy their hands. Every single child was interested in the 5-minute story.

It may not be fair to compare storytelling to anything–storytelling triumphs over rocks, (and even paper and scissors!) in the minds of most humans, but the five-minute story was the most developmentally appropriate activity during the tour. It was the most suitable pedagogical activity in terms of the audience and the timing because looking for books–something the kids were all *capable* of doing–was not needed at that time. There was no assignment, no obvious motivation, and no real reason to add meaning to that otherwise mundane task. It was like teaching 10-year-olds the minute details of how to start a car and put it into gear, but knowing full well that the kids must wait 6 more years before they can check the rear view mirror and pull out into traffic. When given the opportunity, the author turned the public library tour on its head and provided 5 minutes of photocopying and drink getting followed by 25 minutes of raucous storytelling. Why? Because the instructional goal was not to teach children how to find books anymore. That would be taught at some future visit when, individually or collectively, children needed to find books *at that moment*. The instructional goal of the new tour was to make the children want to come back to the library, to make the library a fun place to re-visit.

Katz's examination of learning focuses on what we wish to teach children at various stages of school-age development. We don't want to teach them to be engineers just yet, we simply want them to willingly repeat (and have fun doing) the types of learning tasks that will eventually make them want to become engineers or to think like engineers no matter what they wish to study. Although one might debate a large list of potential matches between school-age behaviors that match well with future engineering skills, we propose these four categories of behaviors as especially worthy of consideration: *Thinking in New*

Ways; Finding Out How Things Work; Designing and Building and Disassembling and Rebuilding; Dreaming and Imagining.

Knowledge, Skills, Dispositions, and Feelings

The division of learning into these four categories helps us to formulate instructional activities in any subject, but it is especially helpful when one looks at long term goals in a particular discipline. In writing, for example, one must have a foundation of basic knowledge (the black marks on paper represent letters, the letters form words using these particular rules) and basic skills (cursive writing or keyboarding skills). That knowledge and those skills do not mean everyone becomes a writer although nearly everyone may do some writing. The disposition to be a writer is a separate set of habits or behaviors that have more to do with the choices people make, the things people value, and the values people appreciate in others.

Lilian Katz calls dispositions "habits of mind or tendencies to respond in certain situations in certain ways" (Katz, 1999, p. 3). Curiosity and creativity—two highly useful dispositions in the field of engineering—are dispositions at the heart of most successful learning in young children. Alfie Kohn notes that dispositions are not learned through formal instruction or exhortation. Many important dispositions, including the disposition to learn and to make sense of experience, are in-born in all children—wherever they are born and are raised. Many dispositions that adults want children to acquire or to strengthen . . . are learned primarily from being around people who exhibit them; they are strengthened by being used effectively and by being appreciated rather than rewarded (Kohn, 1993).

Katz and others (Dweck, 1991) point out that in order to strengthen dispositions, children must be provided with opportunities to express the dispositions through behaviors and observe their effect and their effectiveness. Katz and Dweck advise that teachers can strengthen dispositions most readily by setting learning goals rather than performance goals. For example, a teacher who says, "See how much you can find out about something," rather than, "I want to see how well you can do," helps children focus on what they are actually learning rather than an external evaluation of their performance on a given task (Katz, 1999).

A national and state-by-state focus on competencies and learning standards may have relegated Lilian Katz's last of the four learning areas to a back seat these last few years. Feelings, however, especially those feelings learned in school that promote active learning are still critical for academic success. Feelings about school, teachers, learning and other children have a profound influence on school success as do feelings about one's competence, confidence, and security in the school setting.

Of the four learning areas highlighted by Lilian Katz–knowledge, skills, dispositions, and feelings, this article will focus foremost on dispositions or positive learning habits that will translate best into future engineering behavior.

Collaborative Learning

Contemporary research confirms that children learn most effectively when they are engaged in interaction rather than in merely receptive or passive activities (Bruner, 1999; Katz, 1999). The nature of this interaction will shift across the school years, but there are common elements across the years as well. Collaboration in the learning environment works best when there is room for some informality, especially for younger children (Katz, 1999); when there is a competent person giving children the support needed to engage in a task that would be too difficult to do alone (Vygotsky, 1978); and when there is competent (adult) support for the social interaction taking place in the collaboration (Bruner, 1983). The continuum between collaborative learning in school age populations and various learning innovations at the college level is a good one. Collaborative learning, team approaches to learning, and problem-based learning are all current, successful instructional methodologies found in engineering programs in colleges and universities across the nation.

A SAMPLING OF ENGINEERING PROGRAMS AND EXEMPLARY WEB AND CHILDREN'S BOOK RESOURCES FOR K TO 12

Instructional resources are all the resources, both human and material, that are used in the process of providing the students with the knowledge, skills, and attitudes they need to develop into the type of individuals who are able to acquire an enhanced capability to solve the complex problems of modern society as individuals or on teams, an understanding of the interconnectedness of all knowledge, and a deep commitment to lifelong learning. Consequently, the teacher as a facilitator needs to provide the resources for learning that will help students make their own learning meaningful, and to help students develop those habits of mind that will serve them well into adulthood.

There are a huge variety of engineering resources that will serve the learning goals established earlier–suitably meeting children's developmental levels and needs; focusing on dispositions and feelings as much if not more than knowledge and skills; and emphasizing adult-guided collaborative learning. This paper will selectively sample three types of engineering resources that support that type of instruction in the K to 12 classroom–model programs, web resources, and children's books. These three types of resources are divided among the four categories of behaviors identified earlier as good potential matches between K-12 learning and future engineering dispositions–*Thinking in New Ways; Finding Out How Things Work; Designing and Building and Disassembling and Rebuilding; Dreaming and Imagining.*

THINKING IN NEW WAYS

Model School Programs

Engineering, But How? A Course (Principles of Engineering) for High School Students at Madison West High School, Madison, WI

In response to the significant labor shortages of technologically oriented people faced by the United States, high school engineering teacher Alan Gomez developed the course for senior high students. He finds that learning to navigate the road to a solution is just as important as finding the solution itself. The course is organized around concepts, skills, and attitudes necessary for an engineering career and has students work on real-world case studies that resemble the problems they will be solving in an engineering career (Gomez, 2000).

So What Is Engineering? A Course for Secondary Teachers and Counselors at Cal State L.A.

This continuing education course developed by Raymond Landis, education and technology dean at California State University at Los Angeles, had three basic goals: (1) give participants a broad overview of engineering, highlighting its various disciplines and industrial sectors; (2) outline the different aspects of the engineering education process, including the course of study and the desired level of student preparation; and (3) introduce new ways of working with students to promote effective time management, positive peer interaction, and other learning strategies important for success in engineering school. The course was thought to be especially useful in promoting greater awareness about engineering among high school teachers and counselors who often heavily influence students' decisions about what major to choose in college (Gibney, 1998).

Web Resources

NIEHS Kids Pages (primary grades)
http://www.niehs.nih.gov/kids/home.htm

National Institute of Environmental and Health Sciences website includes links to Games and Surprises to test imagination and shape and spatial relationship skills. The *What's Wrong with These Pictures?* section tests the important reasoning ability–"which of these items doesn't belong." It includes many other excellent thinking exercises for younger children.

Greatest Engineering Achievements of the 20th Century (intermediate through high school)
http://www.greatachievements.org/greatachievements/index.html

This site makes an excellent case for the importance of engineers and engineering in our daily lives, sponsored by the National Academy of Engineering.

Discover Engineering (primary and intermediate grades)
http://www.discoverengineering.org/

Discover Engineering is a delightful website for aspiring engineers (and kids who didn't know they were aspiring engineers until they entered the web site!) that features Games, Puzzles, a Scavenger Hunt as well as a Cool Stuff section and an Idea Center for teachers.

Exploratorium, the Museum of Science, Art, and Human Perception (all grades)
http://www.exploratorium.edu/

A creative assortment of science, education, and creative exercises for the mind produced by the San Francisco Exploratorium Museum, including an excellent section on scale and structures (including bridges) complete with activities for the elementary school. One of the best all-purpose thinking/pondering/brain exploration sites on the web.

Children's Books

Black and White, written and illustrated by David Macaulay. Boston: Houghton Mifflin, 1990. (all ages)

Macaulay is probably children's literature's most engineering-friendly illustrator, but this work is unique. Is it one story or four-stories-in-one? Intriguing enough to have won the Caldecott Medal, the United States' highest honor for art in children's literature.

Walter Wick's Optical Tricks by Walter Wick. New York: Scholastic, 1998. (intermediate)

Using mirrors, lighting, shadows, and simple props, Wick's photographs create a series of optical illusions while his text delivers hints about how he tricked the reader's brain.

I Want to Be–an Engineer, created/produced by Stephanie Maze, written by Catherine O'Neill Grace, photographs by Peter Menzel et al. San Diego, CA: Harcourt Brace, 1997. (primary through middle school)

A nice snapshot of the accomplishments of the various branches of engineering and how dependent humans are on those accomplishments.

Mouse Views: What the Class Pet Saw, written and photo-illustrated by Bruce McMillan. New York, NY: Holiday House, 1993. (primary)

Photographic puzzles follow an escaped pet mouse through a school while depicting such common school items as scissors, paper, books, and chalk. Readers are challenged to identify the objects as seen from the mouse's point of view.

Round Trip, written and illustrated by Ann Jonas. New York: Greenwillow, 1983. (all ages)

One of the best picture books ever created to help children think in new ways. Jonas's black and white illustrations and text record the sights on a day trip to the city and back home again to the country. The trip to the city is read from front to back and the return trip from back to front, upside down, a remarkable feat for an artist and a delightful treat for all ages. How did she ever do that is an often-heard question.

Starry Messenger: A Book Depicting the Life of a Famous Scientist, Mathematician, Astronomer, Philosopher, Physicist, Galileo Galilei, created and illustrated by Peter Sis. New York: Farrar Straus Giroux, 1996. (primary through intermediate)

Peter Sis, one of the most unorthodox and fascinating designers of children's books describes the life and work of the courageous man who changed the way people saw the galaxy. Was there ever a greater challenge to the beliefs of the day?

FINDING OUT HOW THINGS WORK

School Programs

An Aircraft Design Project for the High School Level, Volusia County (FL) School System

Charles Eastlake, a design engineer in the aircraft industry, originally developed this computer-assisted aircraft design project for two different audiences–for low performance level junior high school students to demonstrate that math and science can be used to do interesting and fun things and for high school students with good academic performance to provide a taste of what engineering design has to offer (Eastlake, 1998).

Web Resources

Try Science (primary and intermediate grades)
http://www.tryscience.org/

Try Science is a gateway to experience the excitement of contemporary science and technology sponsored by a partnership among IBM, the New York Hall of Science (NYHOS), the Association of Science-Technology Centers (ASTC), and over 400 sciences centers worldwide. Try Science has a Technology & Engineering link that includes a Build and Test A Paper Bridge option as well as A-Mazing Robots that can be programmed to pick up toxic waste. Several other science options segue nicely into various fields of engineering.

American Society for Engineering Education (middle school and high school grades)
http://www.asee.org/precollege/html/fun.htm

ASEE's Pre-College site provides a good deal of helpful information about engineering and features an "Engineering Can Be Fun" link with Games, PlaneMath, and Control Your Own Robot on the Moon.

Children's Books

Building, by Philip Wilkinson, photographs by Dave King and Geoff Dann. New York: Alfred A. Knopf, 1995. (intermediate and up)

The art and technique of how buildings are constructed from mud huts to city skyscrapers.

Houses and Homes, by Tim Wood. New York: Viking, 1997. (primary and up)

A look at houses and homes throughout history with peeled back roofs and walls and lots of detail for those who love to wonder what it was like to live way back when.

Incredible Everything and *Incredible Explosions,* written by Richard Platt and illustrated by Stephen Biesty. Dorling Kindersley, 1997 and 1996. (intermediate and up)

Biesty is the master of the cut-away detail and no detail is overlooked. His *Incredible Everything* looks at a truly odd yet fascinating mix of how things are made. Included are milk, matches, soap, pipe organs, racecars, wigs, chocolate bars, mummies, photocopiers and doughnuts. Oh, and the Saturn V rocket and false teeth!

The Way Things Work, written and illustrated by David Macaulay. Boston: Houghton Mifflin, 1988. (intermediate and up)

Macaulay's wonderful tribute to technology and engineering is that rare double success–a respected reference work and a browser's joy in one remarkable volume. Dorling Kindersley published the newest CD version in 1999.

DESIGNING AND BUILDING
AND DISASSEMBLING AND REBUILDING

School Programs

Children's Engineering: The Use of Design Briefs

Sigmon reminds the reader about one of Jerome Bruner's best-known hypotheses, "any subject can be taught effectively in some intellectually honest way to any child at any stage of development" (Sigmon, 1997, p.14). Although often misinterpreted, Sigmon correctly points out that the key to Bruner's statement is not that anything can be taught to a child of any age, but that if it is to be taught effectively it should be in an "intellectually honest way." Following that premise, Sigmon examines whether very young children (3- to 6-years-old) can be introduced to design briefs as a way to solve problems. Cognizant of the developmental abilities of this age group and mindful of their typical ways of learning, the author makes a good case for the types of dispositional learning this activity utilizes. She writes: "The learning activities are hands-on and evoke both critical thinking and creative activity. Children learn concepts through firsthand practice, trial and error, and through exploration and discovery. Students involved in technology education are planning, designing, building, evaluating, working together to solve problems–technology education is real life, practical, and not to mention, fun!" (Sigmon, 1997, p. 16).

Girls in Engineering, Math, and Science (GEMS),
Minneapolis Public School System

For four months, two teams of 5th and 6th grade girls met weekly to build Lego robots using Lego Dacta computer software. The project director writes: "At first I heard, 'M.J., help us!' After they realized they would be relying on each other, however, their comments changed to 'Look, we did it!'" Building skills increased as did team success and individual confidence. The girls presented their accomplishments at a conference and at the Science Museum of Minnesota (Savaiano, 1999). Note: Another successful project-based learning experience was conducted at the Lu Sutton Elementary School in Novato, CA. The Design Technology Exhibit challenges students with a design problem and provides them with a chance to share their solution with others through a public forum (Edwards, 1996).

Turning Students into Inventors: Active Learning Modules for Secondary Students

Gorman et al. describe a high school-based series of learning modules that generate student excitement in the process of invention before (as the authors state) the students are turned off by the heavy load of required courses for majors in science and engineering. "This excitement can carry some students into careers in science and technology and help all students recognize ways science and technology have changed the world." A nice example of adult-guided collaborative learning (Gorman, Plucker, and Callahan, 1998).

Additional Models for Engineering Design in Secondary Schools

Engineering design projects utilizing group learning, team learning, and collaborative learning have been demonstrated effectively in secondary schools around the country. Examples include:

- The Southeastern Council for Minorities in Engineering (SECME), a partnership among seven southeastern U.S. universities, regional schools, and local industry, that run the eponymous Science, Engineering, Communications, Mathematics Enrichment (SECME!) program. "[Y]ou won't find students in a SECME classroom sitting quietly in rows, eyes cast down into books or cast upwards–and glazing as a teacher lectures. SECME kids work in teams collaborating on projects, and their teachers roll up their sleeves and get down to work with them" (Hamilton, 1997).
- Southridge High School in Washington where teacher Jim Hendrick's subject is the hard-core, hands-on class called engineering technology, in which students are expected to cross the curriculum to complete his ingenious and multifaceted projects. "What I want to see is kids starting businesses and creating jobs for others," says Hendricks (Lee, 1997).
- LBJ High School in Austin, Texas takes authentic assessment to new heights when students design a solution to a technology problem and are graded on whether they meet the specifications for the job, just as any technology firm would be assessed. Designed by science teacher Jackson Pace and vocational-technical instructor Anthony Bertucci, the course

exposes students to real-life problems and a very real approach to how evaluation takes place in the business world (Phillips, 1998).

Web Resources

NASA Kids (primary and intermediate grades)
http://kids.msfc.nasa.gov/

"Projects & Games" (mazes, games and hunts), "Make it Yourself" and "On-line Activities" are just a few of the features that encourage thinking about designs and the construction of objects (including paper airplanes).

Be a Spacecraft Engineer (intermediate grades through high school)
http://stardust.jpl.nasa.gov/education/jason/index2.html

Users are invited to design a spacecraft that will protect the NASA space station Stardust from orbital debris. Interactive site includes mission training (lessons, graded questions and answers) mission briefing (more in-depth lessons) and user design (users build modified spacecraft online by dragging parts provided into a Word document or offline, on paper or with a computer graphics program). Design page includes questions to aid design analysis.

Lego's Robotics Network (middle school and high school)
http://www.Lego.com/robotics/

Lego's Robotics Network provides a 3D Simulator for future engineers to engage in robotics missions designed to navigate and solve problems in the virtual world. Note: Lego's reputation for providing children with excellent building materials (and the requisite brain stimulation to use them) is well deserved. These seemingly simple pieces of toy building blocks that children use to construct different designs, simple forms, and complex systems, are remarkable in their stimulation of play and imagination. When children play with Legos they are actively involved, and they think actively about the process of design and invention. They have the freedom of experimenting with possible structures and possible solutions to problems encountered during the process of design and construction. Legos enable children to learn vital scientific and engineering concepts in the act of play, one of the truly powerful tools of learning.

Lego Media (all ages)
http://www.Legomedia.com/12k/html/f.asp?go=games/racers/racersmain.html

Lego Media delivers more construction strategies and activities for manipulating, building, and "unbuilding" with the Lego pieces.

Children's Books

Changes, Changes, by Pat Hutchins. New York: Macmillin, 1971. (primary)

Two wooden dolls use their imaginations to rearrange their block world to serve their immediate needs.

The Elements of Pop-Up: A Pop-Up Book for Aspiring Paper Engineers, by David A. Carter and James Diaz. New York: Simon & Schuster, 1999. (intermediate and up)

A beautifully crafted and designed how-to on all the classic pop-up book tricks–a tour-de-force of a tribute.

Castle (1977); *Cathedral: The Story of Its Construction* (1973); *City* (1974); *Mill* (1983); *Pyramid* (1975); *Underground* (1976), written and illustrated by David Macaulay. Boston: Houghton Mifflin. (intermediate and up)

David Macaulay is the champion of building design books for children–no one does it better. Now also well known for his PBS television specials on buildings, Macaulay has inspired many a young engineer and architect. Two additional titles worth singling out are Unbuilding (1980), Macaulay's clever fictional account of the dismantling and removal of the Empire State Building and *Building the Book Cathedral* (1999), a 25th anniversary tribute to Macaulay's first book done by the only man capable of throwing a party of this magnitude–David Macaulay, himself.

Paris Underground, by Tamara Hovey. New York: Orchard Books, 1991. (intermediate)

A detailed and fascinating look at the building of the Paris subway.

DREAMING AND IMAGINING

School Programs

Seabase America, an educational simulation project, Copley-Fairlawn High School, Copley, OH

Carolyn Staudt, with a cast of hundreds, ran one of the most remarkable real world experiments ever held in conjunction with a school. From April 23 to 29, 1994, one hundred children entered a geodesic dome complex in Ohio for an experiment in engineered living that ran for 137 hours. Seabase America, meant to simulate the deep waters off of the Florida keys, was populated and run entirely by the students with support from hundreds of additional students (and many excited parents). The students inside were on their own–no teachers allowed. If a problem developed (mechanical, environmental, social, etc.) the aquanauts solved it themselves, on the spot. Seabase was the ultimate collaborative exercise (Raymond, 1995).

Web Resources

Building Surprises: The Architecture of the Weisman Art Museum (elementary through intermediate grades)
http://hudson.acad.umn.edu/surprises/home.html

From sketch to construction, a look at the building of the University of Minnesota's Art Museum featuring a biography of the architect, step-by-step construction of the building, an overview of its interior and surroundings and a using-your-imagination "What do you see" feature.

MIT Media Lab (intermediate through high school grades)
http://www.media.mit.edu/

The Massachusetts Institute of Technology (MIT) Media Lab is very involved in finding creative ways to use new technologies as a tool for innovative ways of thinking, learning, and designing. Their web site is a fascinating place to take a glimpse at the future. The folks at the Media Lab (and all its subdivisions and groups) are building new "tools to think with." The Media Lab also seeks how these tools can help schools enable students to learn to give personal meaning to concepts, knowledge, skills, and actively create new knowledge. Some of their products especially relevant to engineering include the following:

> *Programmable Bricks*–Programmable Bricks are Lego bricks with tiny computers embedded inside them. They are electronic bricks which children use for constructing computational capabilities directly into their Lego constructions.

> *Programmable Beads*–Programmable beads are computational toy jewelry that enable students to think critically and creatively while playing with them. Each bead is a tiny computer with the capability to communicate with the adjacent beads, via inductive coupling, and to change color. These bead-computers encourage children to experiment with patterns and aesthetics. The behavior of each bead-computer has power over the behavior of the other bead-computers in the string of beads.

Stackables–Stackables are blocks that can only communicate with the block next to it. By putting the blocks together an information path is created and a distributed display is formed through the interaction of the individual blocks. The way the blocks are arranged determines the type of display constructed. The idea behind the construction of stackables came from programmable beads. For more information visit: *http://el.www. media.mit.edu/projects/stackables/index.html*.

National Engineers Week (all ages)
http://www.eweek.org/

The annual creative celebration of the invisible profession–February 18 to 24, 2001 marked the 50th anniversary. The web site includes tons of excellent information about careers, activities, engineering wonders, links to interesting web sites, and a special Sightseer's Guide to Engineering that offers the opportunity to note engineering achievements of significance in one's state.

Children's Books

Harold and the Purple Crayon by Crockett Johnson. Johnson. New York: Scholastic, 1955. (young primary)

Young Harold goes for an adventurous walk in the moonlight with his purple crayon which he uses to get himself out of trouble and eventually to find his way home–the ultimate design-as-you-go adventure dream for the very young.

Girls Think of Everything: Stories of Ingenious Inventions by Women, written by Catherine Thimmesh, illustrated by Melissa Sweet. Boston: Houghton Mifflin, 2000. (intermediate)

A tribute to the ingenious inventors of interlocking bricks, signal flares, vacuum canning, folding beds, fire escapes, school desks, drip coffee machines, and chocolate chip cookies, among many other critical everyday products.

The Trek, written and illustrated by Ann Jonas. New York: Greenwillow, 1985. (primary)

A highly imaginative girl forges her way to school, observing and avoiding all the wild animals posing as trees, chimneys, fences, and even fruit.

REFERENCES

Anonymous. 1997. "Finding poetry in prototypes." *Techniques*, 72 (8): 17, November/December.
Bruner, Jerome. 1999. Keynote address. IN: *Global perspectives on early childhood education*, 1999, Washington, D.C.: A workshop sponsored by the Committee on

Early Childhood Pedagogy, Nancy Academy of Sciences, and the National Research Council, 9-18.

Bruner, Jerome S. 1983. *Child's talk: Learning to use language.* New York: W.W. Norton.

Dweck, Carol S. 1991. "Self-theories and goals: Their role in motivation, personality, and development." IN: *Nebraska Symposium on Motivation: Vol. 38,* Richard A. Dienstbier (Ed.), 1991, Lincoln: University of Nebraska Press, 199-235.

Eastlake, Charles N. 1998. "An aircraft design project for the high school level." *International Journal of Engineering Education,* 14 (1): 54-59.

Edwards, Don. 1996. "Design technology exhibit." *The Technology Teacher,* 55: 14-16.

Gibney, Kate. 1998. "So what is engineering?" *ASEE Prism,* 7 (7): 14-15.

Gomez, Alan G. 2000. "Engineering, but how?" *The Technology Teacher,* 59 (6): 17-25.

Gorman, Michael E., Plucker, Jonathan A., & Callahan, Carolyn, M. 1998. "Turning students into inventors: Active learning modules for secondary students." *Phi Delta Kappan,* 79 (7): 530-535.

Hamilton, Kendra. 1997. "Mousetrap cars, egg drops and bridge building." *Black Issues in Higher Education,* 14 (10), 22-26.

Katz, Lilian G. 1999. Another *Look at What Young Children Should Be Learning.* ERIC Digest (EDO-PS-99-5)–ERIC Clearinghouse on Elementary and Early Childhood Education, Champaign, IL, (ED430735), 4 pages.

Kohn, Alfie. 1993. *Punished by rewards: The trouble with gold stars, incentive plans, A's, praise, and other bribes.* Boston, MA: Houghton Mifflin (cited in Katz, 1999).

Lee, Mike. 1997. "Mixing it up." *Techniques,* 72 (8): 14-17, November/December.

Phillips, Ione D. 1998. "Engineering a grade." *Techniques,* 73 (3), 18-21, 51.

Raymond, Allen. 1995. "It's called 'The Mother of all Simulations.'" *Teaching Pre K-8,* 25 (5): 36-40.

Ryan, Kevin & Cooper, James M. 1998. *Those who can, teach,* 8th ed. Boston: Houghton Mifflin Co., p. 549.

Savaiano, M.J. 1999. "Teaching about robotics." *Childhood Education,* 76 (1): 32.

Sigmon. Jillian F. 1997. "Children's engineering: the use of design briefs: Creating techno toddlers, preschoolers, and kindergartners." *The Technology Teacher,* 56 (7): 14-16, 21-24.

Vygotsky, Lev S. 1978. *Mind in society, The development of higher psychological processes,* Michael Cole, Vera John-Steiner, Sylvia Scribner, and Ellen Souberman (Eds.). Cambridge: Harvard University Press.

Virtual Engineering Libraries

Jill H. Powell

SUMMARY. There are over 3,200 search engines on the Internet, and with so many choices it is not easy to determine which ones find information needed by engineers. This article presents the features and services of ten virtual libraries (also called annotated web directories) that cover engineering networked resources. Large multidisciplinary sites, such as Google and Yahoo, plus smaller sites focused exclusively on engineering or academic subjects, such as the Edinburgh Engineering Virtual Library (EEVL) and Infomine, are compared. Methods used to make comparisons include searching the databases to see which common and uncommon engineering resources were included. Additionally, some engineering keywords were searched in each index and the number of results are presented. *[Article copies available for a fee from The Haworth Document Delivery Service: 1-800-342-9678. E-mail address: <getinfo@ haworthpressinc.com> Website: <http://www.HaworthPress.com> © 2001 by The Haworth Press, Inc. All rights reserved.]*

Jill H. Powell, BA, MLS, is Reference/Instruction Coordinator, Engineering Library, Carpenter Hall, Cornell University, Ithaca, NY 14853 (E-mail: jhp1@cornell. edu).

[Haworth co-indexing entry note]: "Virtual Engineering Libraries." Powell, Jill H. Co-published simultaneously in *Science & Technology Libraries* (The Haworth Information Press, an imprint of The Haworth Press, Inc.) Vol. 19. No. 3/4, 2001. pp. 105-128; and: *Engineering Libraries: Building Collections and Delivering Services* (ed: Thomas W. Conkling, and Linda R. Musser) The Haworth Information Press, an imprint of The Haworth Press, Inc., 2001, pp. 105-128. Single or multiple copies of this article are available for a fee from The Haworth Document Delivery Service [1-800-342-9678, 9:00 a.m. - 5:00 p.m. (EST). E-mail address: getinfo@haworthpressinc.com].

KEYWORDS. Engineering virtual libraries, Web sites, online searching

INTRODUCTION

According to a recent article in the New York Times (Guernsey 2000), there are more than one billion web pages, and that number doubles once every eight months. There are 3,200 search engines, with Google (recently calling itself "giga google") having the highest number of indexed pages–over 1 billion; with half full-indexed, and half partially-indexed (Google Inc. 2000). Automated search engines are having trouble keeping up with the explosion, hiring enough staff, and dealing with expired links. How do you determine which search engines to recommend and use?

I ask students in my library classes to name their favorite search engines, and the response changes every year. This year Google, DirectHit, and Metacrawler garnered the most number of votes. Nationally, Yahoo is probably the most popular, with between 40-63 million visitors per month, according to Media Metrix, an Internet measurement company (Guernsey 2000). Altavista, HotBot, Metacrawler, Excite–all have their fans.

My personal favorite, Google (*http://www.google.com*), cuts down on irrelevancy by pointing to the most popular web pages. It does this by using the PageRank ranking algorithm that uses information from the number of pages pointing to each page. Google also uses the text in links to a page as descriptors, since the links often contain more accurate descriptions than the original web sites themselves (Lawrence and Giles 1999, p. 118). DirectHit (*http://www. directhit.com*) also uses a popularity-based algorithm, by ranking results for a given query according to the number of times previous users have clicked on a page.

While the big search engines use computer programs to scan the Internet for sites and to do the indexing, many also rely on humans to some extent to mine and catalog the Internet for the best resources. This is done to avoid "tyranny of the majority" since users may not always want to see the most popular sites (Guernsey 2000). Google, Netscape, Lycos, HotBot and others use the Open Directory Project for much of their data. The Open Directory Project (ODP) (*http://dmoz.org/*) may be one of the largest human-created directories with over 27,000 editors and 1.9 million sites (ODP 2000). This widely-successful project may surpass the popularity of the spider-compiled directories, and proves that a combination of humans and computers do the best job.

Besides the big search engines, there are also smaller sites which depend on human indexing called virtual libraries, which have much smaller staffs but do a better job of annotating than ODP. This article will examine some of these libraries (also called annotated directories) which specialize in engineering resources on the Internet.

What exactly is a virtual library? Virtual libraries (as in WWW Virtual Library) are directories on the web that contain collections of resources that librarians have carefully chosen, annotated, and organized in a logical way (Ackermann and Hartman 2000). This includes large sites like Google as well as the smaller sites, such as Ei Village and EEVL. Which ones are the best? Which ones will help library patrons find the information they need? Surprisingly, this article finds the answer to these two questions may be different.

THE EDINBURGH ENGINEERING VIRTUAL LIBRARY (EEVL)
http://www.eevl.ac.uk

The Edinburgh Engineering Virtual Library (EEVL) is probably the largest free specialized guide to Internet resources in engineering. Shown in Figure 1, it links to over 25 specific engineering subject areas, has over 5,000 annotated entries and recently recorded its 4 millionth page view (EEVL 2000a). The entries are classified with subject keywords. Layout is clear and well designed. The editors look for current and high-quality sites, including commercial and nonprofit sites. It was founded in late 1995 because accessing the big search engines (Yahoo, Altavista) was too slow from Europe, and there were no annotations on the entries to indicate what type of information a site provides (MacLeod 1997a).

EEVL is run by a team of information specialists from Heriot-Watt University (4.15 full time equivalents), with input from University of Edinburgh, Cambridge University, Imperial College of Science, Technology and Medicine, Nottingham Trent University, University of Sheffield, Institution of Electrical Engineers, and Cranfield University (Kerr and MacLeod 1997, p. 112). The Institute for Computer Based Learning provides technical input (MacLeod 2000c, p. 59). The subject classification is loosely based on the Engineering Index classification scheme and is located at *http://www.eevl.ac. uk/iis/s11.htm*. Their target audience is the British higher education and research community, professional practicing engineers in the UK, the informed public in UK, and the engineering community outside the UK (MacLeod 2000c, p. 60).

EEVL Categories have many subcategories, for example:

Aeronautics (ALL sub-sections)
  Aeronautics (General)
  Aerodynamics
  Aircraft Design
    . . .

FIGURE 1. Edinburgh Engineering Virtual Library (EEVL)

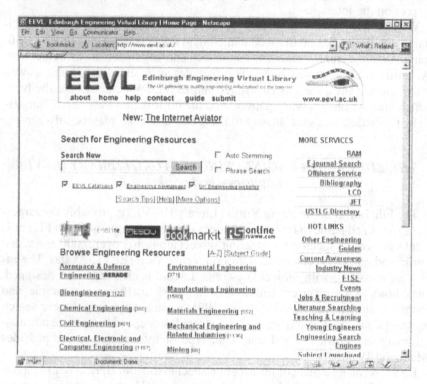

Searching is very flexible. You can browse all entries on Aeronautics, or just some in a sub-section. You may search by keywords and restrict it to a type, such as publishers, or software. Entries are fully annotated, which is very helpful to the reader in determining what information a site provides. Embedded links are included in the annotations, which allow direct access to important sections of a site. A link to "comment on this record" appears at the bottom of each record, which goes to the EEVL webmaster. Figure 2 shows a sample entry from their website.

EEVL has its own search algorithm, which sorts search results based on where query terms are present (title, description, or keywords). EEVL choice sites (those deemed most useful) are listed first. To reduce the number of zero results a user receives, EEVL implemented an automatic stemming option to queries.

The site is extremely well organized. Information about statistics, structure, services, documentation, and help is available at *http://www.eevl.ac.uk/about. html*. Annual reports, press releases, evaluation reports, background papers, and conference papers presented on EEVL are all listed at this address. Any group deciding how to set up a web resources directory on any subject would benefit from viewing their organization.

FIGURE 2. EEVL Record

AIAA Meetings Papers Searchable Citation Database |Non UK|

[Full EEVL Record] http://www.aiaa.org/publications/mp-search.html

The AIAA Meetings Papers Searchable Citation Database, updated quarterly, contains Author, Title, Paper Number, and Conference Date and Location information for papers presented at AIAA meetings from 1992 to today. The database offers a number of search options including: Paper Title or Title Keyword; Author Name or Affiliation; AIAA Paper Number, Year of Paper's Presentation; Database Accession Number; Conference Name Keyword, and, Complete Conference Title. A number of sort options can be used to display records. Copies of AIAA papers can be ordered online as a charged service.
Comment on this record

Over the years new services have been added, including: an e-journal search engine, which searches the full-text of 220 engineering e-journals; ability to restrict searches by 21 types (databases, companies, FAQs, associations, full-text, videos, etc.); an engineering newsgroup searchable archive, covering 125 newsgroups; an engineering resources on the Internet bibliography; and "hot links," which link to current awareness services and news. EEVL recently became the WWW Virtual Library for Engineering, one of the oldest Internet indexes (MacLeod 2000a).

EEVL has developed strong relationships with other engineering information providers. They have loaded specialized databases, such as the Liquid Crystal Database, Recent Advances in Manufacturing (RAM) bibliographic database, Offshore Engineering Information Service, Jet Impingement Database, and University Science and Technology Librarians Group Directory. The reason for this is not to compete with commercial online hosts, but to enable these smaller databases, which might not generate much revenue and be dropped from commercial services, to survive (MacLeod 1997b).

EEVL is funded by the UK Joint Information Systems Committee through the Resource Discovery Network (RDN). These organizations are UK higher education funding councils (MacLeod 1997a). EEVL generates publicity by publishing advertisements and articles in engineering magazines, and distributing calendars, pens, and stickers to its target audience. EEVL also accepts some commercial advertising, but the 1-2 advertisements do not detract from the clean design. Since their JISC funding runs out in 2002, this type of fundraising is prudent and will hopefully assure EEVL's future.

ENGINEERING ELECTRONIC LIBRARY, SWEDEN (EELS)
http://eels.lub.lu.se/

The Swedish Universities of Technology Libraries started the *Engineering Electronic Library, Sweden (EELS)*. It is one third the size of EEVL.

Their best-covered subject areas are: physics, mathematics, energy technology, nuclear technology, light and optical technology, computer science and engineering, general engineering, polar research and cold region technology. EELS arranges entries based on Engineering Information classification terms.

There are 14 information specialists who contribute to the site, each responsible for one or more subject areas. Most do so as an "add-on" to their regular jobs, however, so the site is not as aggressively updated as EEVL's (Jansson 1996). Their fortes are Nordic resources you will not find anywhere else. Entries are annotated and classified. Here is an example of subject hierarchy from their website:

400 Civil Engineering
   400 Civil Engineering, General
   401 Bridges and Tunnels
   402 Buildings and Towers
   403 Urban and Regional Planning and Development
   . . .

Figure 3 shows a sample full entry. The class numbers (100, 200, 400, etc.) may be distracting to those unaccustomed to using them, since most patrons search with keywords or subject headings. It is an interesting example of a different virtual library design.

FIGURE 3. EELS Record

Title
  Byggdok : Swedish Institute of Building Documentation
URL
  http://www.byggdok.se/
Description
     BYGGDOK is the central national body in Sweden providing information and documentation services on building, architecture, regional and urban planning, civil engineering, building services, environmental technology, air and water pollution control and waste management The Institute has a co-ordinating position in the co-operation between Scandinavian countries concerning building documentation by running the "BYGG-FASTIGHET-MILJÖ" database and receiving input to it from the Nordic contributors **Subject Descriptors**
  Civil Engineering, Architecture, Environmental Engineering, Bibliographies, Urban planning, Energy conservation
EI Classifications
  403.1 525.2 903 400 402 450
Resource type
  Bibliographic database
Country of Publishing
  Sweden
Language
  en
Format
  text/html

AVEL (AUSTRALASIAN ENGINEERING VIRTUAL LIBRARY)
http://avel.library.uq.edu.au/

The Australasian Engineering Virtual Library (see Figure 4) is a very well-designed site and has concentrated mostly on Australasian engineering resources. None of the sites from Tables 1 and 2 are included in AVEL, so it is hardly fair to compare it to the other web directories in this article. However, I recommend viewing AVEL because there is much to learn in terms of their superb design and navigation and their excellent coverage of Australasian sources. Using Dublin Core records, AVEL links to web sites, Australian digital theses, jobs, conferences, full-text papers, and an online bookshop, to name a few. Developing partners include the Universities of Queensland, Melbourne, New South Wales, Monash University, Queensland University of Technology, Institution of Engineers, Australia Distributed Systems Technology Centre, and the Centre for Mining Technology and Equipment.

FIGURE 4. AVEL

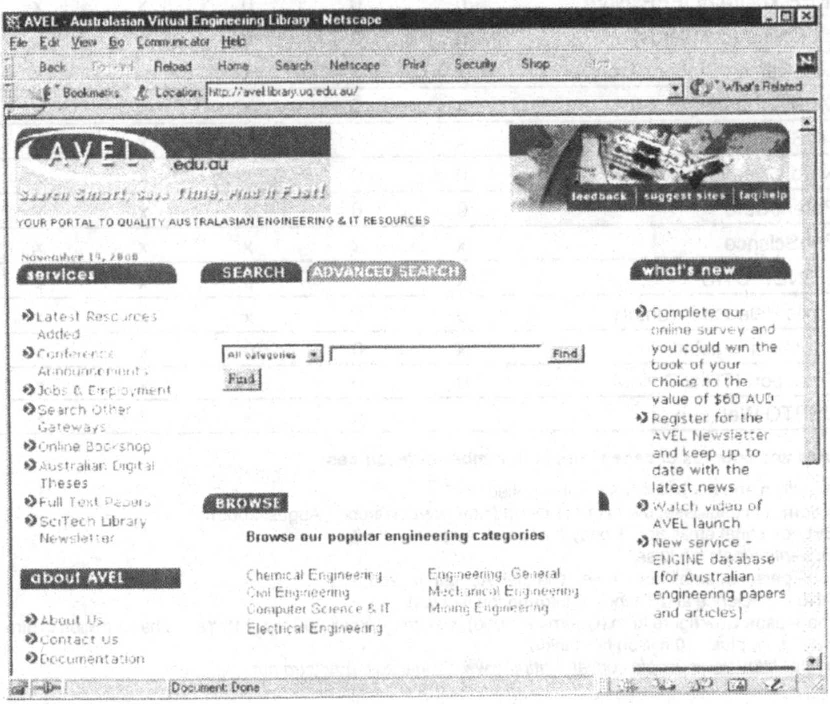

TABLE 1. Types of Resources in Virtual Libraries

	EEVL	EELS	Scout	Yahoo	Google
Total number of resources:	5,858	1,461	9,600	111 m	1 b
Daily page views	6,000	ns	ns	1.7 m	13 m
Score	18	8	14	25	25
AIAA Meetings Papers	x	0	0	x	x
ASCE Civil Engr Database	x	x	x	x	x
ChemExpo.com	x	0	0	x	x
Chip Directory	x	x	0	x	x
Composite Materials Hdbk	x	0	x	x	x
DOE Information Bridge	x	x	x	x	x
Earthquake Engr Abst/Quakeline	0	0	x	x	x
Ei Compendex	x	0	0	x	x
Engineering Case Studies	x	0	0	x	x
Engineering Zone	x	0	0	x	x
European Patent Office	x	x	x	x	x
Icrank.com	x	0	0	x	x
IEEE Xplore or IEEExplore	0	0	0	x	x
ILI Online (standards)	0	0	0	x	x
Los Alamos Preprint Archive	x	x	x	x	x
MEMS Clearinghouse	x	0	x	x	x
NCSTRL	x	x	x	x	x
Nerd's Heaven: Software Dir.	0	0	x	x	x
Patent Café	0	0	x	x	x
PubScience	x	x	x	x	x
STINET–DTIC	x	0	0	x	x
Toxic Release Inventory	0	0	x	x	x
Thomas Register	x	0	0	x	x
Transport (TRIS Online)	0	0	x	x	x
USPTO Web server	x	x	x	x	x

Notes and sources of page views and number of resources:

m = million and b = billion; ns = not supplied
x = item is in database; 0 = item not found (sites were searched August 2000)
EEVL–personal email from Roddy MacLeod, August 2, 2000
EELS–*http://eels.lub.lu.se/*
Scout–personal email from Aimee Glassel, August 22, 2000
STINET = Science and Technical Information Network
Yahoo–page view figure from (Guernsey 2000); size from (Hock 2000, p. 24). Yahoo has 1 million in directory, plus 110 million from Inktomi.
Google–*http://www.google.com* and *http://www.google.com/adv/intro.html*

TABLE 2. Types of Resources in Virtual Libraries, Continued

	Infomine	ODP	Sage	BUBL	Ei
Total number of resources:	22,000	1.9 m	8493	12,000	10,000
Daily page views	5,000-7,000	ns	ns	ns	ns
Score	18	17	16	12	18
AIAA Meetings Papers Database	x	0	x	0	x
ASCE Civil Engr Database	x	x	x	x	x
ChemExpo	x	x	0	x	x
Chip Directory	x	x	0	0	0
Composite Materials Handbook	x	0	0	0	0
DOE Information Bridge	x	0	x	0	x
Earthquake Engr Abst/Quakeline	x	x	x	x	x
Engineering Case Studies	x	0	0	0	x
Ei Compendex	x	0	x	x	x
Engineering Zone	0	x	0	0	0
European Patent Office	x	x	x	x	x
Icrank.com	0	x	x	0	x
IEEE Xplore	0	0	x	0	0
ILI Online (standards)	0	0	x	0	x
Los Alamos Preprint Server	x	x	x	x	x
MEMS Clearinghouse	0	x	0	0	x
NCSTRL (ancestral)	x	x	x	x	x
Nerd's Heaven: Software Dir.	x	0	0	0	0
Patent Café	0	x	0	x	x
PubScience	x	x	x	x	0
STINET–DTIC	x	x	x	0	x
Thomas Register	x	x	x	x	x
Toxic Release Inventory	x	x	x	0	0
Transport (TRIS Online)	0	x	0	x	x
USPTO web server	x	x	x	x	x

Notes and source of page view and number of resource figures:

x = item is in database; 0 = item not found (searched August 2000)
m = million; ns = not supplied
Infomine–(Mitchell 2000a) approximately 2,800/22,000 are science-related
ODP–http://dmoz.org/
Sage–approximately 2975/8493 sites are physical sciences and engineering-related (Hightower 2000)
BUBL–(Williamson 2000) about 320/12,000 resources are engineering-related
Ei–from *http://www.ei.org/engineeringvillage2/quicksearch.html*, from Ei Village 2

BUBL LINK / 5:15 (STRATHCLYDE UNIVERSITY)
http://bubl.ac.uk/link

BUBL is a catalog of resources based on the Dewey Decimal system of classification, and uses subject terms originally based on Library of Congress subject headings, which have been heavily customized and expanded to suit the content of the service. One of the most unique services that BUBL offers is its myriad of search possibilities, which include LC subject, keyword, Dewey class, country, type, and random. The aim is to provide access to key resources for each subject (ideally from 5 to 15), hence the name Bubl Link / 5:15. Every resource in the catalog is annotated, with the last date on which the resource was checked included. The design is quite attractive in that the titles are displayed in one frame with the annotations on the right, scrollable side.

BUBL has over 12,000 resources, 320 of which are engineering-related. A staff of 3 runs BUBL (Williamson 2000). The entire collection of links is verified using link-checking software each month, combined with a program of manual updating, to ensure that links are valid. Like EEVL, BUBL is funded by JISC.

CANADIAN ENGINEERING NETWORK
http://www.transenco.com/

The Canadian Engineering Network, like AVEL, is focused on serving users in a geographic location, in this case, Canada. It includes links to universities, newsgroups, suppliers, architects, research centers, e-commerce centers, software, and more. U.S. resources are not well covered, but that is not the focus of the site.

Ei VILLAGE 2
http://www.ei.org

Engineering Information launched Ei Village in 1995. It is a commercial service now owned by Elsevier. Ei Village has several subscription options available, including the subsets Ei Engineering Village, Ei Paper Village, and Ei Computing Village. A 5-day free trial is available.

In September 2000 Ei Village 2 debuted, with an improved interface. A single search query delivers not only web resources, but can be re-executed to deliver standards, patents, handbooks, and full-text journal articles (see Figure 5). Ei Village 2 includes 5 independently-searchable databases, including Ei Compendex, Website Abstracts for evaluated websites; US Patent Office for patent information, CRC Press Handbooks, and Industry Specs and Standards

FIGURE 5. Ei Village 2

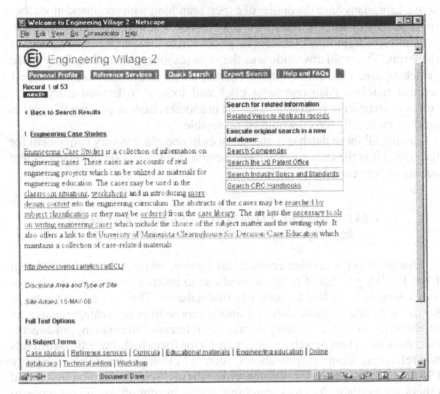

from CSSInfo. Compendex alone abstracts 5,000 journals and conferences. Ei Village subscribers also have access to document delivery, three API EnCompass oil and gas databases, 466 full-text Elsevier journals, Elsevier Advanced Technology (magazines, business newsletters, reference manuals, handbooks and market research reports), and more.

Ei has done research to find out what engineers need and enlarged the scope of their coverage to enable networking and also include non-engineering information engineers require, such as business and travel sites. Some of their services are "Ask a Librarian"; "Ask an Engineer"; which leads to a list of 10 senior engineers; Network of Experts, which is a database of 15,000 professionals also known as the Community of Science; current awareness services; and personal archive space for saved queries.

The entries include more in-paragraph hypertext links than most other engineering Internet indexes. They use the Ei controlled vocabulary and these subject-categories at the bottom of each entry are linked so one can browse further.

Searching is fairly easy in Ei Village 2 and the layout of entries is very organized. Librarians have the option of expert searching with command mode, but most users will want to use quick search. There is no need to use Boolean, proximity, or stemming operators in quick search; the search engine autostems the terms. The main downside was the slow response time, particularly when switching databases. Perhaps that is because the service is still new; Ei Village 1 seemed quicker. Also one must click and look at a detailed entry (on a subpage) in order to re-execute a search in another database; having this option on the first results screen would be preferable.

Having all these databases and services in one place is very convenient for users, and it is the only site reviewed here to include this set of databases and access to peer experts.

DIALOG SELECT OPEN ACCESS ENGINEERING
http://openaccess.dialog.com/tech/index.html

Dialog Select is another commercial service, which, while quite different from Ei Village, should be mentioned due to its enormous coverage. It provides access, for a fee, to hundreds of databases. The user doesn't need to know the database names, they can simply choose from such categories as tech research, tech news, company information, reference information, and directories. Annotated lists of web resources aren't the focus here, but article abstracts and references from many databases, such as Ei Compendex, Global Mobility (SAE), INSPEC, Metadex, Computer Databases, Science Citation Index, and others, are available for cross searching. If your institution doesn't subscribe to these databases, subscribing to this service would enhance your access enormously to technical (and nontechnical) information.

INTERNET CONNECTIONS FOR ENGINEERING
AND ENGINEERING RESOURCE GUIDES
http://www.englib.cornell.edu/ice and http://www.englib.cornell.edu/erg

ICE (Internet Connections for Engineering) was started by librarians at Cornell University in 1994 as a site for listing Engineering resources on the Internet. It has many of the resources listed in Table 1 by category with very short annotations. However, since the search engine, Ice-Pick, is under development, ICE is not as advanced as most of the search engines reviewed here. Approximately half of the ICE resources have been added to the Cornell Library catalog, which has its own search interface. Funding for ICE ended after one year, so efforts to maintain it have been with existing staff–a situation similar to EELS. Realizing that it couldn't compete with the other virtual libraries, and not wishing to duplicate their efforts, the decision was made to launch and focus on a new service that lists the core sites for each field–Engineering Research Guides (*http://www.englib.cornell.edu/erg*) (seen in Figure 6).

These guides, designed for Cornell users, list resources (including many commercial databases) specific to 14 engineering fields as well as key sites useful for all engineering fields (such as standards, dissertations, and patents). Categories are journal indexes, full-text resources, technical reports, and major societies and organizations, and entries are fully annotated. The guides point to many subscription-based sites Cornell provides access to, such as IEEE Xplore, Web of Science, and Ei Compendex, as well as free services such as ASCE Civil Engineering Database and NCSTRL (National Computer

FIGURE 6. Engineering Research Guides (Cornell University)

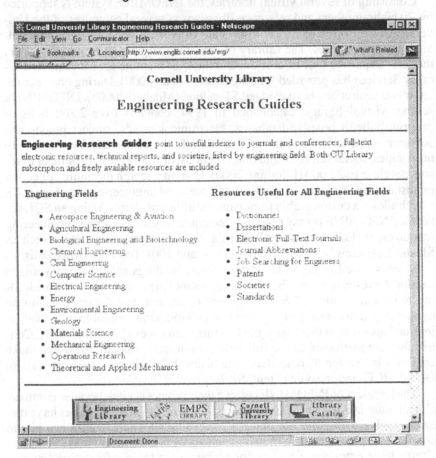

Science Technical Reports Library). A searchable index is almost completed. These guides are much easier to maintain than ICE since they do not attempt to cover all the engineering resources on the Internet, but instead are focused on databases used for academic research that are available for users.

INFOMINE:
SCHOLARLY INTERNET RESOURCE COLLECTIONS:
PHYSICAL SCIENCES, ENGINEERING,
COMPUTING AND MATH
http://infomine.ucr.edu/search/physcisearch.phtml

Consisting of several virtual libraries, the INFOMINE system is supported by 20 bibliographers and reference librarians at several campuses of the University of California and California State University systems, plus other colleges and universities. The Library of the University of California, Riverside, the U.S. Department of Education, and the U.S. Institute for Museum and Library Services has provided funding (Mason et al. 2000). During one year the latter two institutions have pledged $1 million (Mitchell 2000b). INFOMINE's science virtual library, established in 1994, contains over 2,800 links to preprint bulletin boards, databases, electronic journals, subject guides and software repositories, and more. INFOMINE is a cooperative venture, with humanities, social sciences, and other sections, bringing the total number of resources close to 22,000 (Infomine 2000). One may search the entire collection or just a specialty, such as physical sciences and engineering (see Figure 7), which allows a customizable front page for different clients. Using an SQL database, INFOMINE is easy to search and contains some American engineering resources, including hi-tech California ones, that other sites lack, such as Slashdot, Andrew's MacTCP Drive-Thru, and 1001 Top Free Downloads.

Selectors use Library of Congress subject headings and other keywords to catalog the resources, and the search results include a nice link to "terms leading to related resources." Search tips are prominent, and the site contains an explanation to their subject organization and indexing terminology, as well as instructions on how to classify the Infomine resources (Infomine 1996). Contributors are instructed to use full sentences in their annotations, which make entries a bit easier to read than sites using brief phrases, such as Yahoo, Google, ICE, and to some extent, Sage.

The founders of INFOMINE believe their project is vital, because commercial sites are lacking in several respects. They feel commercial sites have distracting advertising, are not objective in listing sites (companies can pay to change their placement to a more advantageous position), and may eventually charge large amounts of money for access to the same information libraries have compiled and made freely available. Smaller institutions will then be unable to afford access. Another reason for the academic virtual library: which

FIGURE 7. INFOMINE

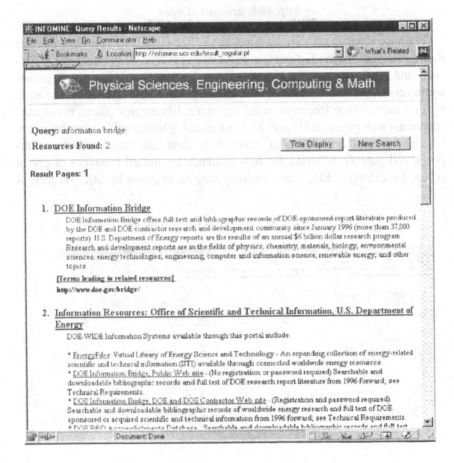

would most scholarly users prefer, an annotated entry made by Yahoo in 2 minutes or one classified by INFOMINE in 25 minutes? (Mitchell 2000b).

One of the most exciting aspects of INFOMINE is its goal to combine both the expert-based virtual library with the current smart crawler/classification-based search engine technology used by Google. This hybrid system should showcase the best of both worlds of the manually-built virtual library (known for link quality and precision) and the Google-type search engine (known for recall and reach). Staff are also working on machine-assisted resource description and collection maintenance systems to reduce costly labor time. By the end of 2000 when this system is expected to be implemented, there should be several hundred thousand searchable resources in INFOMINE (Mitchell 2000a). INFOMINE actively seeks participating members from other academic virtual libraries.

SAGE, UNIVERSITY OF CALIFORNIA, SAN DIEGO
http://libraries.ucsd.edu/

The Sage database of web resources at UC San Diego is a well-designed site that has straightforward searching and ranked results. Originating at the Science and Engineering Library and now grown to 40 contributors, Sage has some very useful and innovative features (Hightower 2000). Results are sorted by information type (recommended resources, library web pages, e-journals, additional web pages, see Figure 8). Besides using linked subject headings, entries also include type labels such as people, data sets, pictures, technical reports, and patents, allowing for new searches to find all instances of patent types, for example. Electronic journals may be browsed by title, subject, or

FIGURE 8. SAGE (University of California, San Diego)

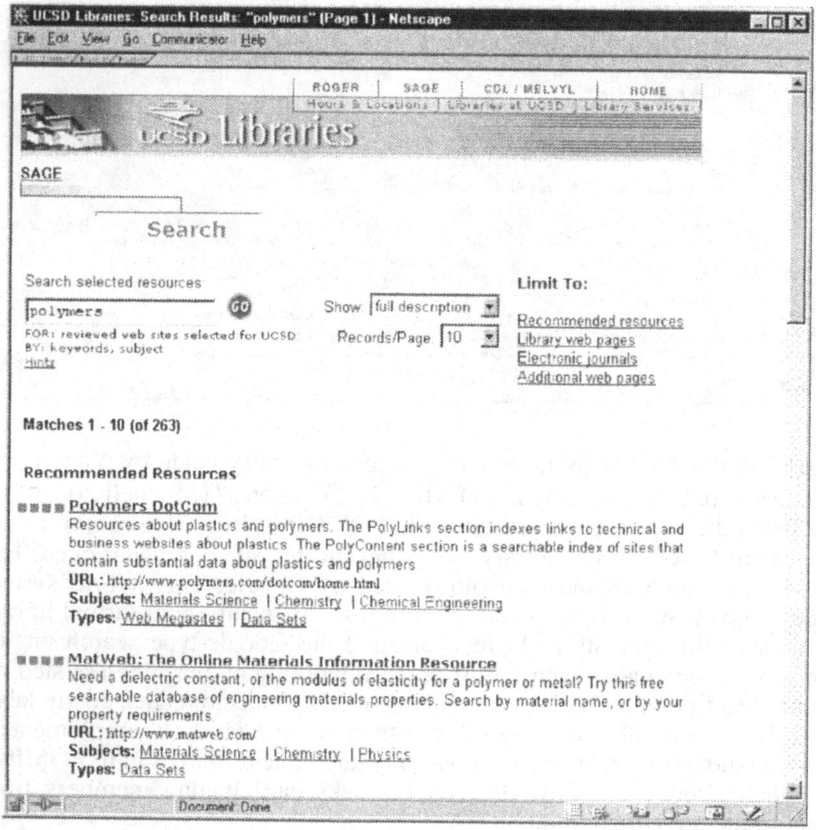

searched by keyword, and these are searched directly from an exported file from the library catalog and brought into Sage.

One of the most interesting aspects of Sage is the flexibility with which its spider indexes. Selectors choose how far down the indexing should go for a particular URL. For example, they may choose just one page (for amazon. com) or the home page plus 2 more subpages (for nasa.gov). This broadens the range of indexing for sites like NASA, which has many installations, projects, and potentially interesting keywords that are hard to include when classifying. Amazon itself would be indexed to the first page only, since it would be categorized as a books-in-print type resource and further subpage indexing would not be necessary (Hightower 2000).

The search results on Sage have noticeably shorter annotations than some of the other libraries. Shorter annotations can be advantageous since they allow you to scan the search results faster, but some added value to the user also decreases. The ranked results, square icons, and clean design do make it visually attractive for the user to scan the results. Both Infomine and Sage go one step further to blend commercial and free web sites into one database, clearly indicating when you need to be affiliated with the institution to access the resource. This is very beneficial to the user, since they will have fewer places to search for their information needs.

SCOUT REPORT SIGNPOST
http://www.signpost.org/signpost

The Scout Report is sponsored by the Internet Scout Project, located in the Computer Sciences Department of the University of Wisconsin-Madison, and was funded by the National Science Foundation. It began as "Net-happenings" in 1993 and become the Scout Report in 1994 (Calcari 1997, p. 11). Its mission is to assist in the advancement of resource discovery on the Internet. It is well-known in most academic circles as highlighting high-quality web sites of interest to the research and education communities every Friday (bimonthly for science and engineering, social sciences, and business and economics). Their main web site, called Scout Report Signpost, has a searchable listing of four years' worth of the Scout Reports. Unlike most big search engines, The Scout Report is highly selective in its choice of Internet resources, using such criteria as richness of content, authoritativeness, and effective presentation. Further information about their criteria is presented in the Library Hi Tech article (Calcari 1997) and on their web page at *http://scout.cs.wisc.edu/scout/ report/criteria.html*.

The Info Scout finds unusual and useful sites, such as "Nerd's Heaven: The Software Directory." About one-third of their resources are given broad Library of Congress class numbers and specific subject headings. Dublin Core is also used. You can do quick, advanced searches, browse by LC subject headings, or LC class. Short descriptions, followed by links to the richer annotations, are included for each item. Figure 9 shows a sample listing (edited for space).

The Internet Scout Project is unique in that staff actually track nearly 150 mailing lists and newsgroups, and major news sources such as the New York Times and National Public Radio for announcements of new Internet services. They also receive many submissions. They mail this list of 15-20 annotated sites weekly to 30,000 users, who forwarded it to many others. The subject-specific guides have approximately 45,000 subscribers, forwarded to an estimated 100,000 people (Robinson 1999). This figure does not include those who access their web site to view the reports. The Scout Project, like EEVL, does ongoing monitoring of its databases, user surveys, and analyzing search logs in order to implement more sophisticated ways of searching and retrieval. It uses Excite as its search engine.

Over the years the Scout Project has added new features, including Network Newsletter, New-List (e-mail lists), and K-12 Newsletter. These projects now operate under Classroom Connect at *http://listserv.classroom.com/archives*. These current awareness services are all moderated by Gleason Sackmann, who started Net-happenings in 1993 (Calcari 1997, p. 13).

The Scout Report initially received a 3-year grant from the National Science Foundation, and it expired in 2000. Typically NSF projects are only funded for two to three years. This loss of funds has reduced the staff to 2 persons. They are soliciting feedback on how the public uses the Scout Report so they can secure funding from government, philanthropic, and commercial sources. They have received 3,000 responses so far, with examples of how the annotated Scout Reports have been indispensable in their work. Many subscribers articulate how the Scout Report differs from other search engines in

FIGURE 9. Scout Report Signpost Record

Nerd's Heaven--Software directory

Nerd's Heaven is the directory of software directories. Dozens of Internet software repositories are organized under subject headings: general software, Internet, operating systems, mathematics, Windows- and Macintosh-specific, user groups, commercial vendors, and religious software. Each directory is annotated, so that the user knows what to expect before deciding whether to visit.
http://boole.stanford.edu/nerdsheaven.html
Scout Report Date of Review: 7/5/96
Date URL Last Verified: 1/19/2000
Signpost Full Description: Available
Subject Headings:
    Computer software -- Computer network resources -- Directories
    Free computer software -- Computer network resources -- Directories
    Operating systems (Computers) -- Computer network resources -- Directories
    Shareware (Computer software) -- Computer network resources - Directories
LC Classification: QA

that it informs them about sites outside their immediate field that turn out to be very useful (Robinson 1999).

The *Sci-Tech Library Newsletter* by Stephanie Bianchi is similar to the Scout Report and is worthy of mention here. It is a free biweekly current awareness service with reviews of many new Internet resources of interest to scientists and engineers. Funded and produced by the National Science Foundation Library, the newsletters can be viewed at *http://www-sul.stanford. edu/depts/swain/nsflibnews/*.

YAHOO, GOOGLE AND THE OTHER BEHEMOTHS

Yahoo and other large computer-generated web sites should not be overlooked. They can be superb in locating information and indexing millions of web pages. For example, searching AIAA in Yahoo locates the many chapters of AIAA as well as the main organization. These are not listed in the human-indexed sites, nor would it make sense to include them all. Yahoo classifies resources in a rigid hierarchy, which can also be helpful. For example, one can find a list of virtual engineering libraries under yahoo/science/engineering/web directories.

Finding the full-text of several articles needed for this paper was a simple process when typing their titles in Google. They were not found through EEVL's search box. However, EEVL does point to them through a wonderful subpage entitled "Engineering Resources on the Internet: A Bibliography" (EEVL 2000c), meaning this information is available through EEVL, but not necessarily in a way a patron would expect. A computer science faculty member reports that he used to rely on Inspec but now uses Google almost exclusively to find articles, because in his field, articles always appear on the web first (Arms 2000, p. 3).

OCLC markets their virtual library, NetFirst, in a unique way. On the results pages in Inspec and other FirstSearch databases they have a button called "web search" which will re-execute a user's search query. The web search button leads to NetFirst, which is a virtual library created by OCLC librarians, covering all subjects and is around 120,000 annotated records. It is only available to FirstSearch customers. This is a clever idea that shows researchers what specialized virtual libraries have to offer. Perhaps including a prominent button to Google from our virtual libraries, or some other search engine even larger than NetFirst, would help the researcher complete the search for information.

COMPARISON

In an attempt to compare the various engineering Internet search engines, I compiled a list of 25 resources in the field of engineering and searched all 10 web directories to see if they were listed (see Tables 1 and 2). Some of the resources chosen are prominent; others important but perhaps less well known (ChemExpo gives chemical profiles with production, supply, demand, and price figures that are used by chemical engineering students; iCrank.com is a mechanical engineering portal that lists vendors, reference data, job listings, software, etc.). The table by no means constitutes a core or complete list, but it gives an indication of how comprehensive the web directories and search engines are.

Tables 1 and 2 show my results. X indicates it was found; 0 indicates it was not. The top scoring sites were Google and Yahoo, which had every single resource. Ei Village, EEVL, and Infomine were tied next at 18; ODP and Sage were very close behind at 17 and 16, respectively. All these services provide much better annotations than Yahoo or Google. The last sites were Scout Report, Bubl, and EELS. Scout and BUBL scored lower because they are general subject directories, and the staff must spend time covering non-engineering sources as well, and hence they miss some of the lesser-known engineering resources and commercial databases. This broad subject coverage can be very useful for those seeking information outside the scope of engineering fields. Bubl and EELS do not carry U.S. web sites as comprehensively as the others do, but their specialties in UK and Nordic resources make them tops in their area.

Some web directories may exclude commercial sites which charge or require registration. But excluding commercial sites, or putting them on another web page, is often confusing to patrons. It is difficult for the patron to distinguish between what is free and what isn't, especially if the institution has a subscription to the site in question. IEEE Xplore (which I tried also as IEEExplore) was only listed in Yahoo, Google, and Sage, and while new, it is a critical full-text resource for electrical engineering journals, and the name is what patrons remember when trying to locate it.

Tables 1 and 2 show the type of sites included in the directories, but they do not indicate quantity by subject, and this is another factor to be included when making comparisons. Therefore, Table 3 shows total number of resources in each database based on an engineering keyword search. Not surprisingly, Google (followed by Yahoo and ODP) wins based on quantity alone. Few of those entries will have high-quality annotations. Sage, Ei Village 2, EEVL, and Infomine are next in terms of highest number of resources retrieved. Sage may be retrieving high numbers due to its new spidering capabilities (described elsewhere in this article). This is similar in ranking to the scoring in Tables 1 and 2. The Scout Report was not included because they do not tally number of search results.

TABLE 3. Numbers of Resources Retrieved by Keyword (Sites Searched August 2000)

keyword	EEVL	EELS	Yahoo	Google	Infomine	ODP	Sage	BUBL	Ei
biomaterials	24	1	28	40,700	2	14	38	0	49
mechanics	139	40	404	1,090,000	47	534	2,467	9	428
nanotechnology	36	1	67	82,300	4	108	54	5	36
java	42	3	2,335	10,400,000	69	3,680	971	10	226
polymers	68	5	634	269,000	24	236	263	2	350
structural	160	31	1,224	1,790,000	56	898	4,601	4	717

ENGINEERING PORTALS

Besides engineering virtual libraries, there are engineering portals, or sites that provide access to a broad array of resources and services, such as e-mail, forums, search engines, online product catalogs, and job information. Examples include iCrank.com, already mentioned, i-engineering.com at *http://www.i-engineering. com*, and EngineerSupply at *http://www.engineersupply.com*. While less research-oriented than engineering virtual libraries, they are important sites for engineers in obtaining information. Roddy MacLeod covers many more portals in his "Panorama of Engineering Portals" article (MacLeod 2000b).

CONCLUSION

This article has examined some of the best search engines to engineering resources on the Internet. Which ones are the best? EEVL stands out, in my opinion, as one of the best with the largest number of engineering-related entries, quality annotations, greatest number of services, creativity, documentation, organization, and maintenance. Since it caters to UK engineers and is slower than the other sites from the U.S., Americans and Canadians should not rely on it alone and bookmarking (and cataloging) Infomine and Sage is highly recommended. Besides their excellent engineering coverage, Infomine and Sage also have the benefit of covering non-engineering subject areas and they have faster response time for North American customers. They appear to be well-funded, with enthusiastic staff and advanced indexing capabilities.

Which directories will help library patrons find the information they need? This depends on the question, of course, but probably the largest search engines which index the most number of web pages, such as Google and others which also have the largest staff and popularity-ranking algorithms. They return results that are close, if no exact matches are found, unlike most smaller directories. Which would patrons rather receive, no hits or hits that come close? Probably the latter. EEVL has tried to overcome the zero-hit problem by expanding the service to cross-searching newsgroups and other engineering resources, which is a good step. The downside is that these are still all-engineering resources and defining engineering in these days of cross-disciplinary research is difficult.

Ei Village had a very high hit rate from the table, and with the debut of Ei Village 2, has a competitive design and navigation features. The U.S. Patents and Trademark Office and Thomas Register databases that Ei packages with their services are actually free to all users. A cost-conscious institution could survive using the free virtual libraries and subscribe only to Compendex and other databases they need. However for businesses which can afford it and do not have many journal or database subscriptions, having all the Ei services (including the network of experts) in one place would be very convenient.

One could argue Google is best the vast majority of the time not only because it usually finds the information needed but also because it is extremely fast. Also, engineering librarians and engineers do need to look for information that is not engineering-related. What if one needs to know the publisher of some medical journal? Typing the name in Google brings up their home page in seconds. Typing the name in most engineering virtual libraries brings zero hits. But what if one wants to find obscure information instead of the most popular? Google assumes that the majority of people who know what they are doing outnumber the minority, but is that always true?

EEVL, Infomine, and the other annotated directories are extremely useful for finding sites, particularly when working from a broader perspective (what type of circuit design information can one find on the Internet) or when trying to find local information (EEVL for UK; EELS for Swedish sites). Their annotations are very useful in describing a site and this saves time for a user. However, it would be hard to rely exclusively on handcrafted annotated directories in serving the engineering community since they will never cover the Internet to the extent the large sites do. Rather, one should use a variety of search engines and directories, keeping in mind their various strengths.

In an ideal world every institution would have access to a national network of annotated links to scholarly resources, plus customization (or brand naming) available so users would know their institutions are paying hundreds of thousands of dollars in subscription costs for certain databases. We need to continue to educate administrators and the public that access to many databases is not free and that libraries need funding to pay for them. In addition, individual universities need the ability to tailor descriptions for their own users (for example, indicating trial subscriptions, group log-in procedures, etc.) to make using the virtual libraries as seamless as possible.

There are many other virtual libraries (such as Northern Lights and Ask Jeeves) containing engineering resources that I did not have space to review. I encourage your comments and suggestions about these sites and any others.

REFERENCES

Ackermann, Ernest, and Karen Hartman. 2000. *Glossary to "Searching & researching on the Internet & the World Wide Web"* 30 June [cited 25 July 2000]. http://www.webliminal.com/search/glossary.htm#virtual libraries.

Arms, William Y. 2000. Automated digital libraries, how effectively can computers be used for the skilled tasks of professional librarianship? *D-Lib Magazine* 6 (7/8). http://www.dlib.org/dlib/july00/arms/07arms.html.

Calcari, Susan. 1997. The Internet Scout Report. *Library Hi Tech* 15 (3-4):11-18.

EEVL. 2000a. Edinburgh Engineering Virtual Library, http://www.eevl.ac.uk.

EEVL. 2000b. Edinburgh Engineering Virtual Library Basics. http://www.eevl.ac.uk/basics.html.

EEVL. 2000c. *Engineering resources on the Internet: a bibliography* [cited July 2000]. http://www.eevl.ac.uk/bibliog.html.

Google, Inc. 2000. *Google launches world's largest search engine* [press release]. 26 June [cited 25 July 2000]. http://www.google.com/pressrel/pressrelease26.html.

Granum, Geir, and Phil Barker. 2000. An EASIER way to search online engineering resources. *Online Information Review* 24 (1):78-82.

Guernsey, Lisa. 2000. The search engine as cyborg. *New York Times*, 29 June, G1.

Hightower, Christy. 2000. 9 August. Personal phone call.

Hock, Randolph. 2000. Web search engines: more features and commands. *Online* 24 (3):17-26.

Infomine. 1996. *Infomine for indexers*, 4 January [cited 25 July 2000]. http://infomine.ucr.edu/participants/help/indexers.html.

Infomine. 2000. *Infomine: database content information* April [cited 25 July 2000]. http://infomine.ucr.edu/welcome/introduction/description.html.

Jansson, K. 1996. Indexed and quality assessed Internet resources–the EELS project. *Tidskrift for Documentation* 51 (1-2):14-20.

Kerr, Linda, and Roddy MacLeod. 1997. EEVL an Internet gateway for engineers. *Library Hi Tech* 15 (3-4):110-118.

Lawrence, Steve, and C. Lee Giles. 1999. Searching the web: general and scientific information access. *IEEE Communications Magazine* 37 (1):116-122.

MacLeod, Roddy. 1997a. EEVL: past, present and future. *Electronic Library* 15 (4):279-286.

MacLeod, Roddy. 1997b. *The lesser of two EEVLs*. [cited 23 August 2001. http://www.ariadne.ac.uk/issue12/eevl/intro.html.

MacLeod, Roddy. 2000a. *EEVL News Nuggets* (23) 22 March. http://www.ariadne.ac.uk/issue23/eevl/ariadne2.html.

MacLeod, Roddy. 2000b. Panorama of engineering portals. *Free Print*. (66). 6 July. http://www.freepint.co.uk/issues/060700.htm#tips.

MacLeod, Roddy. 2000c. Promoting a subject gateway: a case study from EEVL. *Online Information Review* 24 (1):59-63.

Mason, Julie, Steve Mitchell, Margaret Mooney, Lynne Reasoner, and Carlos Rodriguez. 2000. INFOMINE: promising directions in virtual library development. *First Monday* 5 (6). http://firstmonday.org/issues/issue5_6/mason/index.html.

Mitchell, Steve. 2000a. 8 August. Personal e-mail communication.

Mitchell, Steve. 2000b. Five views of INFOMINE. http://infomine.ucr.edu/participants/ netgain/INFOMINE5views.html.
ODP. 2000. *Open Directory Project* [cited 31 July 2000]. http://dmoz.org/.
Robinson, George. 1999. Web Search Group Loses Grant. *New York Times*, 11 November, G4.
Williamson, Andrew. 2000, 4 August. Personal e-mail communication.

The Digital Engineering Library:
Current Technologies and Challenges

William H. Mischo

SUMMARY. Digital library technologies and implementations continue to evolve. Current digital library work emphasizes the integration of distributed information resources with support services that facilitate locating, retrieving, and organizing this digital content. The present working environment of distributed information resources includes publisher full-text repositories, periodical index databases, Web-based resources, Web portals and search engines, and local information resources such as the online catalog and custom databases. Emerging full-text standards, such as the extensible Markup Language (XML), and linking standards, such as the Digital Object Identifier (DOI), will play prominent roles in improving access and retrieval over these heterogeneous digital resources. Several example applications, developed at the Grainger Engineering Library at the University of Illinois at Urbana-Champaign and designed to address digital library integration and linking issues, are presented. The role of libraries and librarians in this evolving digital library environment is discussed. *[Article copies available for a fee from The Haworth Document Delivery Service: 1-800-342-9678. E-mail address: <getinfo@haworthpressinc.com> Website: <http://www.HaworthPress.com> © 2001 by The Haworth Press, Inc. All rights reserved.]*

William H. Mischo, BA, MA, is Engineering Librarian and Professor of Library Administration, Grainger Engineering Library Information Center, University of Illinois at Urbana-Champaign, Urbana, IL 61801.

[Haworth co-indexing entry note]: "The Digital Engineering Library: Current Technologies and Challenges." Mischo, William H. Co-published simultaneously in *Science & Technology Libraries* (The Haworth Information Press, an imprint of The Haworth Press, Inc.) Vol. 19. No. 3/4, 2001, pp. 129-145; and: *Engineering Libraries: Building Collections and Delivering Services* (ed: Thomas W. Conkling, and Linda R. Musser) The Haworth Information Press, an imprint of The Haworth Press, Inc., 2001, pp. 129-145. Single or multiple copies of this article are available for a fee from The Haworth Document Delivery Service [1-800-342-9678, 9:00 a.m. - 5:00 p.m. (EST). E-mail address: getinfo@haworthpressinc.com].

KEYWORDS. Digital libraries, XML, Digital Object Identifier (DOI), full-text resources, information retrieval technologies, simultaneous search techniques

INTRODUCTION

The terms 'digital library,' 'virtual library,' and 'electronic library' have all been used interchangeably as labels for the broad concept of a network-based library that would provide access to information resources without regard to place and time (Saunders, 1999). The digital library concept has proven to be both difficult to clearly define (Borgman, 2000) and to put into practice (Feldman, 1999). The term has been broadly applied to disparate entities such as library online catalogs, various overarching portal Web sites, and specific collections of publisher full-text materials. The delivery of full-text information to the desktop has been one of the mainstays of the digital library (Saunders, 1999). However, the present proliferation of heterogeneous, discrete publisher full-text sites serves as an example of the disjointed and chaotic nature of today's digital collections and has been held up as the antithesis of an organized digital library (Crawford, 1998).

It has become clear that a digital library must emphasize the integration of disparate digital collections with support services that facilitate locating, retrieving, and organizing information resources (Guenther, 2000a). This must be done within an environment of evolving information technologies. The widespread adoption of the World Wide Web and the acceptance of the Web browser as the universal interface have changed the digital library forever (Guenther, 2000b). First attempts to build digital libraries have typically aimed at integrating resources such as the local online catalog, a consortium catalog, locally loaded and licensed online abstracting and indexing (A & I) services (periodical index databases), locally produced digital resources, collections of related Web sites, and licensed full-text collections. It is this attempted seamless integration of local and remote collections with custom user services and access tools that leads us toward the holy grail of the digital library. Complicating the quest is the fact that the technologies being utilized within digital collections and services continue to evolve.

This paper will describe the evolving information technologies, standards, and services that are being used in the construction of digital libraries. These technologies will provide enhanced access to and navigation through digital resources. The integration and linking of full-text articles and associated metadata information will play an important role in the future of digital libraries. The examples and elements of a digital library will be presented within the framework of the working engineering library environment at the Grainger Engineering Library Information Center at the University of Illinois at Urbana-Champaign (UIUC). It is important to note, however, that these information standards and tools are being widely utilized in other library environments in the sciences, social sciences, and humanities.

FULL-TEXT DOCUMENTS

A major factor in the dramatic growth of electronic full-text has been the rapid evolution of electronic document representation and transmission technologies and standards. At the present time publishers and vendors are using a number of these full-text representation techniques. Even a specific publisher may be utilizing more than one technique. It is extremely important that information professionals be familiar with these various representation technologies in order to better understand and evaluate vendor and publisher systems and also to better evaluate the capabilities of a particular system. It is also important for librarians to be aware of trends in this important area. The diagram presented in Figure 1 shows a continuum of digital full-text representation technologies—from the least flexible to the most robust. Again, all these technologies are presently in use. This snapshot of full-text technologies is presented from a Web-based view that emphasizes the information environment in which libraries operate.

With the introduction of the personal computer and word processing software, large stores of machine-readable documents became available in text format. However, the word processing formats, like the typesetting formats, are typically non-standard proprietary text formats that do not lend themselves to the Web environment.

With the growth of the Web, HTML became the standard format for document representation. The Web has made available well over one billion documents in mark-up format. HTML is an instance of SGML (Standard Generalized Markup Language) that is dominated by presentation tags, as opposed to content description tags. HTML (at least up to HTML 4.0) does not readily lend itself to scholarly publishing activities in the sciences and engineering because of its inability to properly display entities, mathematical symbols, and other special characters. HTML remains a presentation-oriented language with inadequate semantic tools for the effective indexing and fine-granularity searching required for effective retrieval in an academic journal environment. HTML

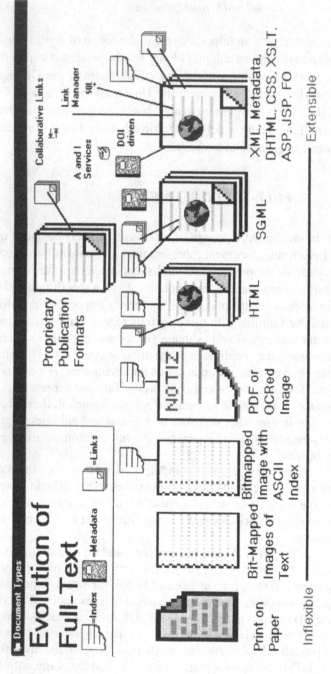

FIGURE 1

introduced the use of client-side scripting and Cascading Style Sheets (CSS) for dynamic display and rendering.

The portable document formats, particularly Adobe's PDF format, allow more sophisticated document representation and are Web-enabled. The PDF format is based on printer output that preserves the appearance of the document, but lacks the capability of fine-granularity markup of the structure and content of the document and does not easily allow the introduction of dynamic links. Because PDF lacks the ability to identify the content of the document at a detailed level, searching can only be done on a fairly crude level, and an ASCII metadata database of document surrogates is typically required for search and retrieval.

PDF is presently the most commonly used full-text format and it is easily adaptable to the Web environment through the use of the Adobe Acrobat plug-in viewer. However, the PDF format does not provide a fine-granularity full-text searching capability and does not easily provide the dynamic linking from a bibliography item to its available full-text and multimedia capabilities that the Web provides. The use of a standard mark-up language, such as SGML and now XML, is regarded as the best and most sophisticated model available for representing text objects in terms of both effective searching and dynamic rendering (DeRose et al., 1997).

SGML was first approved as an international information interchange standard in 1986. SGML is a data description language, or metalanguage, that allows fine-granularity markup of the content and structure of a document. However, SGML is inherently complex to use and, more importantly, has never been adequately adapted to the Web environment.

The eXtensible Markup Language (XML) was established as a Web standard in 1998. It strives to make the best features of SGML more accessible to Web authors and publishers. XML is, in fact, a distinguished subset of SGML specifically designed for Web applications. XML also lends itself to the creation and expression of bibliographic material metadata (Miller, 2000). The major Web and commercial publishers have embraced XML as central to their industry's future (Weil, 2000). In addition to its document representation role, XML is also being touted for its database and transport functionality. But XML is not a database management system, only a support technology for database functions.

The XML specification, like SGML before it, does not explicitly deal with the presentation of content, nor does it inherently supply mechanisms to transform one form of document object into another. These issues must be addressed through the use of CSS and XML Style Language Transformative (XSLT). It is necessary to use these three technologies in tandem—along with a robust database management system—in order to create powerful and robust text applications on the Web. The major strength of XML lies in its ability to expose the deep content and structure of a document for more effective and efficient retrieval.

It is clear that a rich markup format such as XML will become the standard for document representation and delivery in the Web environment. The XML technologies now place us on the threshold of the true 'smart document' shown on the far right in Figure 1. The smart document will provide dynamic linking between full-text documents, A & I Services, and other Web sites. It will be able to 'seek out' similar or linked documents and it will provide collaborative mechanisms for author, reader, and librarian communication.

ILLINOIS DIGITAL LIBRARY TESTBED

Many libraries and information centers have been actively engaged in digital library activities. The University of Illinois at Urbana-Champaign (UIUC) Library has been actively investigating digital library technologies in conjunction with several grant programs. The primary focus of the UIUC program–headquartered in the Grainger Engineering Library–has been the construction and deployment of a multi-publisher full-text Testbed comprised of physics and engineering journal articles (Mischo and Cole, 2000; Schatz et al., 1999). This Testbed has provided UIUC researchers experience with a number of full-text and associated technologies. A demonstration version of the Testbed can be examined at http://dli.grainger.uiuc.edu/.

At the onset of the Testbed project in 1994, there was some speculation that libraries could simply replace print-based journal subscriptions with their corresponding electronic versions with little difficulty. The Testbed project closely examined the issues connected with the local migration from a print-based journal environment to an Internet-based model. The processing, indexing, storage, and retrieval system requirements for the local deployment of a full-text journal repository require significant investments in software, hardware, and staff resources. Building and maintaining this infrastructure is costly. In addition, the economic and licensing issues that surround negotiations with publishers and vendors are daunting.

An important concern of the Testbed group has been in exploring effective retrieval models for the evolving Web-based scholarly communication system. The retrieval and display of full-text journal literature in an Internet environment poses a number of issues for both publishers and libraries. It has now become commonplace for publishers to provide Web-based access to the electronic versions of their serial publications. For academic libraries, support for this publisher-based online journal environment introduces new levels of budgeting concerns and involves an examination of library collection policies, user access mechanisms, networking capabilities, archiving policies, availability of proper equipment, and a greater awareness of requisite licensing agreements.

Libraries have not historically structured information retrieval services around discrete publisher repository collections. There is a need for creative mechanisms to provide effective search and retrieval across the burgeoning number of distributed, heterogeneous publisher repositories. Within the project, the Testbed team devised a distributed repository model that 'federates' or connects the individual publisher repositories of full-text documents. In the Grainger Testbed model, these distributed repositories are federated by the extraction of normalized metadata, index, and link data from the heterogeneous full-text of the different publishers. This model addresses the challenge of providing standardized and consistent search capabilities across the distributed repositories.

The Testbed funding has allowed the Grainger Library Team to explore the application of various digital library technologies within the context of a working engineering library. While the Testbed has focused on full-text journal article and repository technologies, the complementary tasks of providing mechanisms for integrating full-text, A & I Service records, the online catalog, and other Web resources have also been addressed.

In particular, the Illinois Testbed project studied various mechanisms for linking cited articles to their digital full-text representations. The Testbed team looked at both proprietary and standards-based methods for what is called reference linking or link management. One standard destined to play an increasingly important role in reference linking is the Digital Object Identifier or DOI (Atkins et al., 2000). The DOI was developed by the Association of American Publishers and is now managed by the International DOI Foundation (IDF). The DOI is both a unique identifier of a piece of digital content and a system to provide access to the content. Norman Paskin of the IDF has termed the DOI 'the ISBN for the 21st century.'

The DOI is an open standard for digital content identification. It is basically an alphanumeric string that uniquely identifies a digital object. It consists of the Registration Agency Prefix, the publisher prefix, and a publisher-assigned suffix or ID. An example DOI then would be:

10.1063/S0036915.

The suffix can be a dumb number or be based on information such as journal name, volume, issue, and page. The DOI and a URL that point to the digital object are then registered with the IDF. For example:

10.1063/S0036915 | http://www.pubsite.org/j1/apr99/art1.pdf.

The DOI standard is being utilized by the CrossRef service, a reference linking service operated by over 50 of the world's leading scientific, technical, and medical publishers (http://www.crossref.org/).

One concern about the DOI as implemented in the CrossRef protocols has to do with an issue referred to as the 'appropriate copy' or 'localization in reference linking' problem. This problem centers around the ability of a library to provide its users with access to alternate legal copies of e-journal articles that are either mounted locally or available through a license agreement with a third party aggregator, such as OCLC or Ebsco. See http://www/niso.org/linkrept.html and http://www/niso.org/DLFarch.html for detailed explanations of the problem. For example, a library or consortium that has loaded Elsevier ScienceDirect locally would want to refer their users to the local copy of the article and not direct users to the publisher site which will be stored in the Handle server. It is important to note that the CrossRef deposited link will direct users to the article or secondary link at the publisher site, where, in some cases, the user will be turned away because the local library may not be licensed to access the repository directly.

A joint project involving CrossRef, the Digital Library Federation (DLF), the IDF, the Grainger Testbed, OhioLink, Elsevier, and the Los Alamos Physics Preprint Archive to test local reference linking issues is underway.

DISTRIBUTED REPOSITORY ISSUES

A diagram of the overlapping information environment in which academic science/engineering librarians operate is shown in Figure 2. This diagram models the major information nodes that are involved in establishing digital library information services. It also suggests the multiple relationships between the major players in today's information environment.

The primary nodes, as represented in Figure 2, are:

- Discrete full-text publisher repositories;
- Local and remote A & I Service databases such as INSPEC and Compendex;
- Local and remote handbook and databook sites;
- The campus and consortial online catalog (OPAC);
- Web portals (such as EEVL);
- Web resources (such as patent and standard databases);
- Web Search Engines (Google, Altavista, Yahoo); and
- Local databases (such as e-journal lists, serial check-in data, custom reference materials).

This model provides us with a context to discuss the issues that need to be addressed in the current information environment and a starting point to examine the user needs that demand our attention.

FIGURE 2

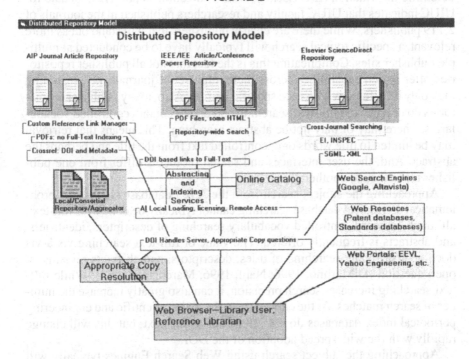

The goal of the digital library in this environment is to provide users with gateway and navigation tools that present an integrated view of these resources and provide seamless and transparent access and linking mechanisms. The realization of this goal presents some exciting challenges. It requires some tools and technologies that are not yet fully developed. It is also safe to say that, as the technology changes and evolves, this goal will require continuous fine-tuning and modification. If there is one constant in the world of information technology, it is that things will change rapidly.

For example, let's examine the needs of the user wishing to locate and read full-text articles in a specific subject area. When approaching this information need from the publisher repository side, there is a major problem in comprehensively identifying the appropriate publisher repository to search. In the present publisher-centric scholarly communication model in which we function, users are often forced to perform a series of separate searches over a number of heterogeneous and discrete publisher full-text repositories (Mischo and Cole, 2000). Cross-repository searching tools are not generally available and within the publisher repository cross-title searching may not even be possible.

The faculty and researchers of a major university publish in a wide variety of journals from a large number of publishers. For example, the Institute for

Scientific Information (ISI) Local Journal Utilization Report (LJUR) data for UIUC indicates that UIUC faculty and researchers published in the journals of 2,119 publishers. While there are clearly publishers that will stand out as more relevant, a specific topical search will typically have to be conducted at multiple publisher sites. Complicating this is the fact that not all publisher repositories offer search capabilities across their entire set of journals, i.e., searching may only be offered within each specific journal. Also, many publisher repositories do not provide full-text searching and do not contain controlled vocabularies. There may or may not be abstract searching. This means that retrieval may be limited to title words or uncontrolled text from the title and perhaps the abstract. And, the user interfaces and search protocols differ from one publisher repository to another.

Approaching the subject search from the A & I database side is also problematical. Periodical database searching cannot typically be done over full-text, although keyword controlled vocabulary searching of descriptors, identifiers, and abstracts is frequently offered. The value of full-text searching vis-à-vis document surrogate searching of titles, descriptors, and abstracts remains an open question (DiMartino, 1996; Nahl, 1996; Marchiano, 1993). While full-text searching increases search precision, it can also greatly increase the number of search matches. At the current time, the major scientific and engineering periodical index databases do not offer links to full-text, but this will change rapidly with the widespread adoption of the DOI.

Approaching the subject search using Web Search Engines typically will yield some useful results, but much of the relevant academic level data is behind firewalls in the locally licensed publisher repositories and periodical index databases.

There are several access and linking capabilities that need to be added to the available systems in order to provide enhanced search and discovery. Within periodical index databases, links to full-text articles need to be made available via DOIs and, at the same time, mechanisms to address the Appropriate Copy problem need to be put in place. And, similar to the situation with the distributed publisher repositories, methods to choose the most relevant database need to be offered and/or simultaneous database search mechanisms need to be introduced. For simultaneous search mechanisms to be truly successful, there will need to be vocabulary switching via mappings from the controlled vocabulary of one database into the controlled vocabulary of others.

The role of the periodical index databases in the distributed repository model is not completely clear. A & I Services continue to provide cross-publisher search and discovery within the broad disciplines that they cover, but this is a function that the publishers are also interested in becoming involved. The services that can provide full-text via links to publisher repositories or aggregator sites will remain relevant. We may see a consolidation of A & I

Services. Additional competition for the commercial A & I Services will come from the PubMed, PubRef, and PubSCIENCE initiatives.

The publishers could also move toward taking over the function presently being provided by the A & I Services by establishing federations of publisher metadata databases. These multi-publisher metadata repositories would provide a central search capability and links to full-text at individual repositories. Indeed, in the CrossRef project minimal-level metadata with each DOI is being deposited in a central database. However, this metadata includes only the first author, the article title, and citation information. The CrossRef metadata does not include controlled vocabulary terms, abstract, and additional authors. This will seriously compromise its retrieval capabilities.

Clearly, there are access issues that need to be addressed and better integration of resources within the different nodes is needed. We remain some time away from a transparent one-stop-shopping information environment for users.

INTEGRATION TECHNOLOGIES AND EXAMPLES

Numerous techniques to integrate digital library resources have been developed. This section will present several examples developed at the Grainger Library. The Grainger Library has developed tools that attempt to address various aspects of this problem. The Grainger top-level menu, available at http://www.library.uiuc.edu/grainger/top.asp, attempts to collect and describe relevant resources in a portal-like Web site. A 'Help Getting Started' module has been developed to assist users with resource selection.

The "Currently Received" and "All Grainger Journals" resources integrate journal check-in information with full-text links and issue tables of contents dynamically extracted from ISI's Current Contents database. Providing expanded access to full-text journals and conference series from within different contexts is an important service challenge. At the Grainger Library, availability and link information for approximately 700 engineering related full-text journals is maintained in a custom database. This e-journals database is used to provide links to full-text journals within the Grainger Table of Contents (TOC) service, the Grainger Journals database, and several custom databases that provide holdings of the ACM and IEEE publications. In particular, the full-text links provided in the TOC service have proven to be extremely popular and useful. Over 240 College of Engineering faculty members presently receive e-mail containing weekly tables of contents from the Ovid Current Contents and Engineering Index databases. These e-mail contents are created from a faculty member's custom search profile. The tables of contents include, when available, a link to the full-text of the journal or conference.

The new book titles database is linked with the faculty interest database to provide automatic e-mail notification to faculty of new book and series titles that match their stored research interest profile.

Services that provide the capability of simultaneous searching of heterogeneous resources will play an important role in the integration of digital library components. The Grainger Library has developed and is testing a Web-based module that provides simultaneous searching of the online catalog, and the Compendex, INSPEC, Current Contents, and Applied Science and Technology databases. The search interface is shown in Figure 3. This software takes the user-entered search argument and normalizes it into the format expected by each of the database systems. The searches are performed asynchronously and the numbers of search hits are displayed as they are returned. The short entry results are displayed to the user, via a proxy server, from the actual vendor system that was searched. All subsequent links pass through the proxy server and are redirected to the vendor system with session connection information intact. The search results are in Ovid format, but the proxy server has added links to full-text articles when available. These links to the full-text journal are gener-

FIGURE 3

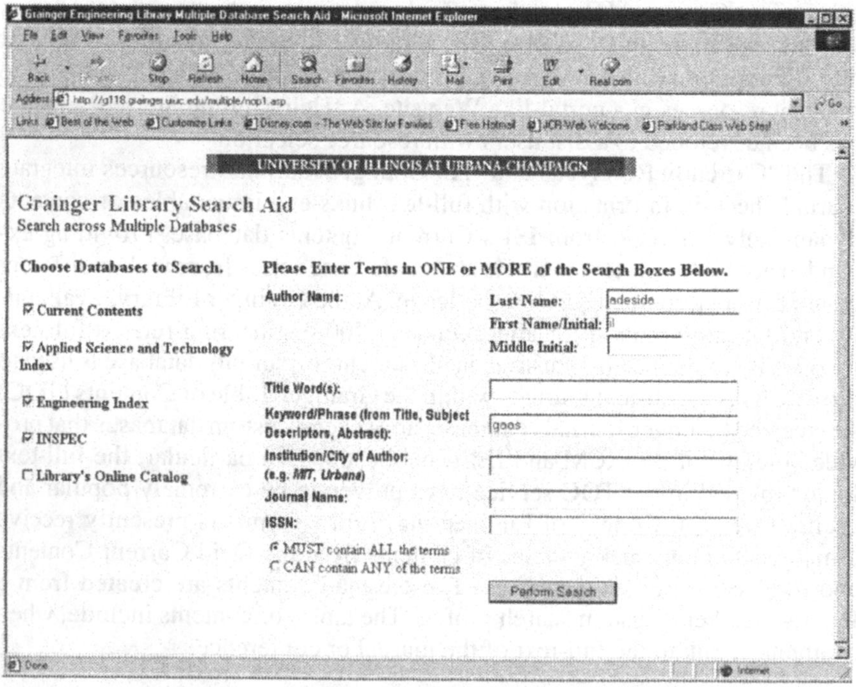

ated from information stored in the custom Grainger e-journals database. The links are programmatically inserted into the value-added Ovid display. Work is being done to add other additional periodical index databases and Web search engines into the Grainger simultaneous search system.

Also being investigated is Searchbot technology, in which terms taken from a user information need statement or search argument will be searched against multiple databases, Web search engines, and portal sites. Information–including citations, full-text articles, Web resources–will be retrieved, distilled, and packaged for presentation to the user. Because the Searchbot will often need to go out and collect, in some cases, a large number of Web pages, this process could take quite some time. In this case, the search result, in the form of a URL, would be e-mailed to the user. The Grainger Searchbot software is an extension of Web harvesting software that was developed to extract faculty research and publication data from individual faculty Web sites and deposit it into the faculty interest database. This Web harvesting software sequentially visits a set of individual URLs stored in a database and extracts and processes the desired text from each given site.

This is an exciting time to be involved in science and engineering librarianship. Engineering libraries continue to make available periodical index and report databases and to license new full-text resources. These include: standards and specification databases from ILI Inc., Information Handling Services, and NSSN; new government databases such as the Transportation Research Information Service system; periodical index databases from Cambridge Scientific such as the Aerospace Database and the Materials Database; and staples such as Compendex and INSPEC. Almost all of the major engineering publishers provide full-text e-journal repositories. This includes societies such as the Association of Computing Machinery, Institute of Electrical and Electronics Engineers/Institution of Electrical Engineers, American Institute of Physics, American Society of Mechanical Engineers, American Institute of Aeronautics and Astronautics, and commercial publishers such as Elsevier, Academic, and Kluwer. Reference handbook and databook repositories are also becoming more common.

Also valuable are Web portals focusing on providing access to engineering resources. The most prominent of these are the Edinburgh Engineering Virtual Library (EEVL) (http://www.eevl.ac.uk/), Yahoo Engineering (http://dir.yahoo.com/Science/Engineering/), Australasian Virtual Engineering Library (http://avel.edu.au/), Engineering Village (http://www.ei.org/ev2/ev2.home), and the GrayLIT Network (http://graylit.osti.gov/). These portals offer rich search capabilities for end users and library reference staff.

It is also sometimes useful to merge information from diverse resources to provide value-added data for users. For example, Grainger serials databases contain usage, citation, and publication information in addition to library holdings records. Publication and citation information from ISI's LJUR data, the ISI JCR (Journal Citation Report) site, the Compendex database, and the INSPEC database have been added to Grainger serial records. The LJUR data, covering 1981 through 1999, provides information on the number of articles written by UIUC authors and articles cited by UIUC authors on an individual journal basis. This data has proven useful for students and faculty who wish to evaluate the relative stature of specific journal titles.

The Grainger Library is also employing Web-based database technologies to provide access to locally generated databases. These databases provide access to resources that are uncataloged (e.g., the UIUC technical report collection and a faculty research interest database) or require expanded access points (e.g., a comprehensive reference collection index) (Mischo and Schlembach, 1999). These Web-accessible databases, which employ Microsoft's Active Server Platform, include: the Grainger Reserves collection database, the Reference Collection database, a New Book Titles database, currently received and retrospective Grainger journals, a multimedia and CD-ROM holdings database, the faculty interest file, the UIUC College of Engineering technical reports collection, a prominent faculty publications database, an Acoustic Emission Collection database, an engineering societies publications database, and frequently asked questions and hard-to-find citations databases.

COMPUTER TECHNOLOGIES

Historically, libraries have utilized whatever computing and information technologies have been available to provide patron services (Crawford, 1999). The continuum of technologies that began with punch cards and batch processing systems passes through networked multiprocessing systems, minicomputer-based turnkey systems, CD-ROM technologies, personal computers, and the Internet. Indeed, the introduction of high-density disk drives such as the IBM 3350 directly led to the success of Dialog, SDC, BRS and the other online database vendors.

Several important information technologies that will dramatically affect digital library services are looming on the horizon. Probably the most pervasive will be the 'ubiquitous computing' technologies. These will be centered on appliance devices such as Palm PCs and Pocket PCs, but will also include next-generation cell phones and pagers. Other portable computing initiatives include the reintroduction of network appliance machines in the form of the SunRays, the legacy free PCs, and other thin desktop machines.

The ubiquitous handheld and portable devices will feature wireless network connectivity and Web access. While there are presently major security and bandwidth issues connected with wireless computing, it is expected that these barriers will fall with the rapid adoption of standards such as Bluetooth and IEEE 802.11 and the standardized instant messaging protocols.

In addition, computing hardware will take advantage of faster processors and increasingly higher bandwidths in order to more broadly utilize multimedia technologies such as streaming video and audio, speech recognition and synthesis, and interactive shared communications and whiteboard technologies.

ROLE OF THE LIBRARY AND LIBRARIANS IN THE DIGITAL LIBRARY ENVIRONMENT

It appears evident that the evolving digital library will continue to emphasize the integration of distributed full-text, secondary, and local information resources. In the short term, this will be accomplished under the aegis of the present publisher-based scholarly communication model and will increasingly utilize Web-based document representation standards such as XML and linking standards such as the DOI. Providing the mechanisms to identify, locate, and obtain information resources in this environment will present a challenge to academic libraries and librarians.

The traditional function of the library has been to collect source materials, organize those materials, and provide effective and efficient access to the materials. While these activities will not change in the digital library environment, the mechanisms and tools to accomplish these functions have changed. The traditional activities have become more distributed and virtual and not confined to a specific place. The universe of available information resources has grown beyond the walls of the modern online catalog.

The overarching question that needs to be addressed in the digital library environment is: How will the support services for the evolving digital library need to change? Clearly, the library has become as much a function as it is a place. It is critical that libraries continue to play an important role in the organization's information infrastructure. At the campus level, this means competing with academic and administrative computing centers, the Chief Information Officer's division, in some cases the School of Library and Information Studies, and departmental workstation laboratories and media centers. Libraries will need to continue to take the lead in licensing full-text and periodical index databases from publishers and vendors for the entire organization.

Another important role that the library and librarians can play is in providing effective gateway and navigation tools for the myriad digital collections in existence at the campus, national, and international levels. Again, this role is related to the library's goal of making digital libraries out of discrete, heterogeneous digital collections. Libraries need to play a leading role in developing the next generation of campus and national information systems. Librarians need to continue to test techniques for simultaneous searching of multiple resources, remote reference and electronic point-of-contact service, remote instruction, software-aided search modification, dynamic linking of resources, and one-stop-shopping approaches.

One of the biggest challenges to the deployment of robust digital libraries is the need to provide remote electronic reference services (Janes et al., 1999). Remote reference services need to be in the form of direct, interactive remote assistance using whiteboard and multimedia technologies and also point-of-contact instructional or assistance software modules. Several of the Grainger local databases have been designed to assist both users and reference staff. These resources include the Grainger reference collection database (with its enhanced access points) and several Web-based bulletin boards used to post and pass timely information, including one maintained by Reference Staff to answer recently asked and class-assignment type questions. A search hints page provides database specific information on truncation symbols, default Boolean operators, and other nuances of specific online resources. A database of frequently asked questions and difficult-to-find citations collected over a number of years is also being made available.

The library also needs to move up on the 'information food chain' of resources consulted by a user with an information need. Studies show that people will first turn to their personal collection, followed by a local colleague, a colleague via e-mail, the Web, and finally the library–in its manifestation as place (i.e., call a reference librarian) and portal site.

There are many questions concerning the traditional role of libraries as archival repositories. As the archival role switches to publishers, a number of questions arise. Can publishers really serve as the sole archivers of their publications? What is the role of libraries in digital archiving? How many mirror sites or backup sites are needed to maintain a true archiving function? Libraries and librarians will be actively involved in determining resolutions to these issues.

Librarians are extremely knowledgeable of the information seeking process employed by users and are often acutely involved with the research and instructional programs in their particular departmental or college's sphere of interest. Librarians need to leverage these skills, in conjunction with their knowledge of information technologies, to forge more effective partnerships with the campus administration, campus organization units, and state and outside agencies.

BIBLIOGRAPHY

Atkins, Helen; Lyons, Catherine; Ratner, Howard; Risher, Carol; Shillum, Chris; Sidman, David; Stevens, Andrew "Reference Linking with DOIs" *D-Lib Magazine* 6, February 2000 (http://www.dlib.org/dlib/february00/02risher.html).
Borgman, Christine L. *From Gutenberg to the Global Information Infrastructure: Access to Information in the Networked World*, Cambridge, MA, MIT Press, 2000, p. 33-52.

Crawford, Gregory A. "Issues for the Digital Library" *Computers in Libraries* 19, May 1999, p. 62-64.

Crawford, Walt "The Danger of the Digital Library" *The Electronic Library* 16, February 1998, p. 28-30.

DeRose, S.J.; Durand, D.G.; Mylonas, E.; Renear, A.H. "What is Text, Really?" *The Journal of Computer Documentation (ACM SIGDOC)* 21, August 1997, p. 1-25. [Reprinted from *Journal of Computing in Higher Education* 1 Winter 1990, p. 3-26.]

DiMartino, Diane J.; Zoe, Lucinda Rhea "End-User Full-Text Searching: Access or Excess?" *Library and Information Science Research* 18, Spring 1996, p. 133-149.

Feldman, Susan E. "Digital Libraries '99" *Information Today* 16, October 1999, p. 1.

Guenther, Kim "The Evolving Digital Library" *Computers in Libraries* 20, February 2000a, p. 48-50.

Guenther, Kim "Designing and Managing Your Digital Library" *Computers in Libraries* 20, January 2000b, p. 34-36.

Janes, Joseph; Carter, David S.; and Memmott, Patricia "Digital Reference Services in Academic Libraries" *Reference and User Services Quarterly* 39, Winter 1999, p. 145-150.

Marchionini, Gary; Dwiggins, Sandra S.; Katz, Andrew "Information Seeking in Full-Text End-User-Oriented Search Systems: The Roles of Domain and Search Expertise" *Library and Information Science Research* 15, Winter 1993, p. 36-69.

Miller, Dick R. "XML: Libraries' Strategic Opportunity" *Library Journal netConnect* Summer 2000 (http://www.ljdigital.com/xml.asp).

Mischo, William.H.; and Cole, Timothy W. "Processing and Access Issues for Full-Text Journals." In *Successes and Failures of the Digital Library Initiative, Proceedings of the 35th Annual Graduate School of Library and Information Studies*, University of Illinois of Urbana-Champaign, 2000, p. 21-40.

Mischo, William H.; Schlembach, Mary C. "Web-Based Access to Locally Developed Databases," *Library Computing* 18, November 1999, p. 51-58.

Nahl, Diane; Tenopir, Carol "Affective and Cognitive Searching Behavior of Novice End-Users of a Full-Text Database" *Journal of the American Society for Information Science* 47, April 1996, p. 276-286.

Saunders, Laverna M. "The Human Element in the Virtual Library" *Library Trends* 47, Spring 1999, p. 771-787.

Schatz, B., Mischo, W.H., Cole, T.W., Bishop, A., Harum, S., Johnson, E., Neumann, L., & Chen, H. "Federated Search of Scientific Literature: A Retrospective on the Illinois Digital Library Project" *IEEE Computer*, 32, 1999, p. 51-60.

Weil, N. "Web Publishers Hinge their Future on XML" *Infoworld* February 11, 2000. (http://www.infoworld.com/articles/hn/xml/00/02/14/000214hnseybold.xml) [24 April 2000].

Building Better Library Web Sites:
State of the Art and Future Trends

Christy Hightower

SUMMARY. Engineering library Web sites are evolving from a static set of html pages to a dynamic database-driven architecture. OCLC's Cooperative Online Resource Catalog (CORC) may be the next step in this evolution. Several key issues in Web design for academic science and engineering Web sites (including the Rapid Evolutionary Development Model, strategies for improving access to content, filtering, zones and automating the production of electronic journal lists) are discussed. Usability testing is highly recommended. It is speculated that the future of reference on the Web may lie with those libraries or other service providers that offer proactive real-time reference using Web collaboration software. *[Article copies available for a fee from The Haworth Document Delivery Service: 1-800-342-9678. E-mail address: <getinfo@haworthpressinc.com> Website: <http://www.HaworthPress.com> © 2001 by The Haworth Press, Inc. All rights reserved.]*

KEYWORDS. Usability testing, user interface, Web site design, effectiveness, database-driven Web sites, Web reference, science library Web sites, Web portals

---

Christy Hightower is Engineering Librarian, University of California Santa Cruz. When this paper was written, she was Engineering Librarian and Web Coordinator at the Science and Engineering Library, University of California San Diego.

[Haworth co-indexing entry note]: "Building Better Library Web Sites: State of the Art and Future Trends." Hightower, Christy. Co-published simultaneously in *Science & Technology Libraries* (The Haworth Information Press, an imprint of The Haworth Press, Inc.) Vol. 19, No. 3/4, 2001, pp. 147-163; and: *Engineering Libraries: Building Collections and Delivering Services* (ed: Thomas W. Conkling, and Linda R. Musser) The Haworth Information Press, an imprint of The Haworth Press, Inc., 2001, pp. 147-163. Single or multiple copies of this article are available for a fee from The Haworth Document Delivery Service [1-800-342-9678, 9:00 a.m. - 5:00 p.m. (EST). E-mail address: getinfo@haworthpressinc.com].

EVOLUTION OF ENGINEERING LIBRARY WEB SITES

Engineering library Web sites seem to be evolving from a static set of html pages to a dynamic database-driven architecture. In the dynamic model the database itself can be created and used locally, or it can be created by one or more institutions working together and accessed as a whole remotely. EEVL, the Edinburgh Engineering Virtual Library (*http://www.eevl.ac.uk/*), and INFOMINE (*http://infomine.ucr.edu/*) are both examples of database-generated remote portals. In the near future you can expect this evolution to move a step further as projects like OCLC's Cooperative Online Resource Catalog (CORC) develop in which the records from many institutions are pooled but each library selects only the records they want for their local database implementation. This evolution is not a moment too soon if libraries expect to keep up with our competitors in the information delivery sector of the Web.

In the early days (beginning around 1993 for the early adopters) engineering libraries constructed local hand-coded Web sites, consisting of static html pages or files. Although the pages were updated from time to time, for any query the user was offered the same pre-determined set of pages from which to select. Viewing the site was like turning pages in a published book. Adding a search engine helped, but the site remained essentially static because the search results were simply pointers to the same pages (with the same unvarying arrangement of resources) that we had been showing patrons before. For instance, a search for "Compendex" might result in a pointer to the static "List of Databases in Engineering" page. Once there you still had to use the "Find in Page" option in your browser to find the Compendex link and annotation. The whole page was the smallest unit of information that could be manipulated and offered for viewing.

Many of these static engineering Web sites were (and still are) quite wonderful in terms of content selection and description. The problem was that as Internet resources multiplied, these sites didn't scale well. Their information delivery wasn't as precise as could be wished, and their maintenance burden was high, especially when links to the same resource were needed on multiple pages.

In the mid 1990's it became evident that moving to a database-driven (or dynamic) Web site could improve the precision of the information delivered, serve interdisciplinary users better, eliminate link redundancy, make the sites easier to update, and provide better maintenance tools for Web page authors.

In a dynamic Web site the links and their annotations reside in a database, thereby making the fields in the record (e.g., each link or annotation) the smallest unit of information that can be served up. The smaller the units, the more precisely targeted the information delivery becomes. The Web pages are created automatically from the database, either on the fly in real-time in response to a user's query, or they can be created ahead of time in response to a pre-determined or frequently asked query. Even when the pages are generated from the database in advance, the links and their metadata remain easier and faster to update than those on traditional static pages. And because the patron can get a page constructed to his specifications, precision increases.

Large virtual libraries like EEVL and INFOMINE were among the first to take advantage of the database-generated architecture. INFOMINE's database was one of the earliest, going live in May of 1994. In addition to reaping the benefits of being dynamic, these large portals also pioneered the use of cooperative resource selection, cataloging and link maintenance by librarians from more than one institution. Sharing the cost of development is very appealing. However, these portals don't integrate seamlessly with your local Web site. When you are in the portal different interface design and navigation schemes apply, and integrated links back to your local Web site, to your locally licensed content, or to local services are not provided. Unfortunately, the local library doesn't have access to the database itself and can't select individual records from the portal. The smallest unit the portals generally offer is an entire Web page of resources.

Individual engineering libraries can make the switch to database-driven Web sites themselves and the trend seems to be to do so, although it does require some programming resources. The University of California San Diego (UCSD) Science & Engineering Library made the change in late 1999. Another approach is to merge the engineering library Web site with that of all the other libraries on campus, to make a single dynamic library "portal." The Libraries at Cornell University did this in January 1998,[1] the University of Washington in Seattle in September 1998, and UCSD in October 2000. The advantages of a local implementation are that the resources in the databases are uniquely tailored to the local clientele (though there are inevitably compromises when the engineering library is part of the larger campus library portal). Furthermore, local database-driven Web sites seamlessly integrate the resources licensed for local use with those available freely on the Internet, while also directing patrons to the appropriate local library services (like interlibrary loan and instruction).

The next evolutionary step is to combine the advantages of a locally created dynamic Web site with the power of cooperative cataloging. A major initiative in this arena is OCLC's Cooperative Online Resource Catalog, or CORC (see *http://www.oclc.org/oclc/corc/about/corc_over.htm*). CORC became available as a billable service from OCLC in July 2000.

CORC extends the OCLC model of shared cataloging to Web resources. Subscribers to CORC use an automated Web interface to point the CORC system to a URL they wish to harvest. The CORC software extracts some basic data from the resource to create a base record. The librarian then edits and completes the record, placing it in the database where it becomes available for use by any other subscriber. The system can use MARC or Dublin Core fields. A full discussion of metadata is beyond the scope of this article, but good starting points are *Introduction to Metadata: Pathways to Digital Information*[2] from the Getty Information Institute (selections are available at *http://www. getty.edu/gri/standard/intormetadata/*) and the IFLANET Metadata Resources Web site at *http://www.ifla.org/II/metadata.htm.*

In addition to tools that alert authors to broken or redirected URLs, CORC also provides linked authority control using the Library of Congress names and subject authority files. Once linked, if the established form changes, CORC will automatically update all the linked resource records. What really makes CORC innovative is the way it makes individual records available to libraries to use as they wish. Using CORC is like having access to the database of a large virtual engineering library so you can take advantage of the description work of the experts (the cataloging) while retaining control of the selection decisions. From the entire database you can select just those records that are the best match for your local community. You can choose to create Web pathfinders that contain only CORC records or you can import the records to fully integrate them into your local system. CORC doesn't yet have all the fields you might want, you lose some functionality (like the authority control updating) when you unlink the records to import them, and the system is not free of charge, so CORC is not nirvana. But it's certainly a very promising evolutionary development.

WEB DESIGN ISSUES FOR ENGINEERING LIBRARIES

There are several excellent books and Web sites on the topic of Web site design (see the recommended reading section at the end of the article). Therefore instead of presenting a general tutorial on the topic, the following is a discussion of several key design issues that are pertinent to academic science and engineering library Web sites. The advantages of the database-generated design over traditional static html pages have already been mentioned. But whether your Web pages are hand-made or machine generated, the following issues are important to consider to ensure that your site achieves maximum usability for research and instruction.

DESIGN DARWINISM
AND RAPID EVOLUTIONARY DEVELOPMENT

In the past when libraries created a new service they had the luxury of an initial phase of slow research and development. Ten years ago OPAC and bibliographic database vendors spent a long time developing their interfaces before bringing them to market, and they solicited the input of responsible librarians to do a great deal of in-house experimentation and interface testing before the products were released to the general public in the library. Once released, expectations were that the products were stable and that significant changes would be few and far between. Extensive instructional materials for these systems could be written and printed on expensive glossy paper because they would not change very often, maybe less frequently than once a year.

Those idyllic days are gone. As James Gleick describes it in his book *Faster: The Acceleration of Just About Everything,*[3] "We're speeding up; our technology is speeding up; our arts and entertainment and the pace of invention and change–it's all speeding up." This is particularly true of the Internet environment, where new features or interface changes are appearing without advance warning all the time, often fueled by the advancing technology that keeps making so much more possible. And rapid development is not limited to the commercial sites like Amazon.com or AltaVista; bibliographic database interfaces and other vendor-supplied library applications also change with increasing rapidity.

Embracing such rapid change is a real paradigm shift for academic libraries. We are accustomed to delivering finished products, not to making public an endless series of draft versions that are continually undergoing incremental change. While librarians may at first feel uncomfortable with this situation it's wise to face up to the fact that the accelerated pace of change appears to be here to stay. The prevalence of freely available information resources and simple-to-use, customer-centered search systems on the Internet has provided our clientele something to compare us to. So unless we are willing to suffer greatly from the comparison, it behooves libraries to develop strategies that enable them to refocus on user needs and change quickly with the times.

On the other hand, rapid Internet development doesn't have to mean completely abandoning the behind-the-scenes development phase and descending into the harsh competitive environment that Jacob Nielsen calls "Design Darwinism":

> ... the Web is developing as we speak, and experiments happen on the open Internet with us all as test subjects–not in a videotaped usability lab. The result is a much harsher Design Darwinism, where ideas crash and burn in public. Eventually, the best design ideas will survive and bad ones will decline because users will abandon poorly designed sites.[4]

There are software development models that streamline user testing without relegating you to "crash and burn in public." Recently, the libraries at the University of California, San Diego have successfully incorporated many aspects of the software development model known as Rapid Evolutionary Development into their Web design workflow. They have also rededicated themselves to usability as the prime directive in their Web design. This approach is very user-centric because it solicits end-user direction and feedback very early in the development process as well as iteratively throughout, yet it's also designed to deliver a product in a short amount of time. This appears to be a successful model from which other libraries could also benefit.

Basically, Rapid Evolutionary Development says

- If it's worth doing, it's worth doing badly (at first)
- The customer often doesn't know what he wants (until he sees it)
- User needs are always changing
- Focus on needs not wants or wishes
- Provide an iterative negotiation between customer and programmer
- Use the 80/20 rule

The overall goal of the Rapid Evolutionary Development approach is to rapidly deliver a working, usable prototype that focuses on the 20% of the functionality that will provide 80% of the benefit. The prototype works poorly at first, but it evolves and develops, growing much as a child does from infancy to adulthood. The prototype is used as the basis for an ongoing dialogue between the programmers and the customer, through iterative usability testing. In the library environment, the customer includes both librarians and end-users. Prototyping is useful because it allows customers to see the functionality and demonstrates what's happening far better than can be explained in the abstract. However, when necessary for expediency "paper prototypes" (printed or hand-drawn screen mock-ups) can be used to test with end-users before committing to a working prototype. The product is made public when an acceptable level of functionality is reached, but it's never "finished." It continues to evolve and change.

The rapid evolutionary design methodology has proven most effective when the processes themselves are subject to ongoing change. It gives you a place to start when initial requirements are vague or future technological advances are imminent but have not yet arrived. It requires a tolerance for ambiguity, and an acceptance of the fact that your product will be continually changing, but it's a workflow ideally suited to the current requirements of the Internet environment. For more information, consult Lowell Jay Arthur's book *Rapid Evolutionary Development: Requirements, Prototyping & Software Creation*.[5]

WHAT TO INCLUDE AND HOW TO MAKE IT FINDABLE

Instant gratification and one-stop shopping is what everyone is after. As Jacob Nielsen puts it "The Web is an attention economy where the ultimate currency is the user's time ... The cost of going to a different Web site is very low, and yet the expected benefit of staying at the current site is not particularly high ... Web content must give immediate benefits to the users or they will allocate their time to other sites."[6] For libraries, our competitors include the likes of Google[SM] (*http://www.google.com*), Amazon.com (*http://www.amazon.com*) and AskJeeves, (*http://www.ask.com/*) and their content and functionality are nipping at our heels.

Aim to improve the integration of print and electronic materials as much as you can. Catalog the best Web sites in your OPAC. And vice versa, export records from your OPAC (for electronic journals or for any other easily identified group of materials that you choose) and create Web pages from those records that are indistinguishable from the rest of your Web site. (For more on this, see the section on e-journals below.) Even in a hand-coded Web site you can choose to include the top five printed reference works for which there is no online equivalent, and write convincing annotations to "sell" their use to your patrons. Otherwise the print resources will languish.

For browsing purposes, point to the big engineering portals, but not as a substitute for your own Web site. As previously mentioned, the large virtual engineering libraries like EEVL and Ei Engineering Village™ (*http://www.ei.org/*) provide a wealth of content, but once your patrons go there they have left your site completely and may thereafter miss access to your locally licensed content or to your local services such as interlibrary loan or document delivery. Unfortunately, there is currently no way to integrate the best or most relevant parts of the engineering portals into your local Web site without duplicating all of their work. Look for future opportunities to encourage sharing of the expensive tasks of selecting quality resources and then creating and maintaining metadata for them in ways that don't compromise your local choice or limit access to local services. (As previously mentioned, CORC is a development to watch in this regard.) Meanwhile, for browsing purposes consider pointing to the specific subsections of the portals that are most relevant to your needs, rather than simply leaving people at their front doors.

For keyword searching an ingenious solution for pointing to portals or to any complex and data-rich site (for example, the Web sites of the National Aeronautics and Space Administration [NASA] or the Department of Energy) is to create your own local full-text index of these sites, to create your own AltaVista so to speak. The SAGE system (*http://libraries.ucsd.edu/*) from the University of California San Diego Libraries does just that.

SAGE is a search engine (powered by the Netscape Compass Server software) that takes as its starting points the URLs of the resources that have been hand-selected by subject specialists for inclusion on the libraries' Web site. (The hand-selected sites are what you see when you browse resources by subject or type. Think of the browsing part of the site as equivalent to Yahoo, while the searching part is analogous to AltaVista.) In the database record for each Web resource the subject specialist has indicated how deep the search engine should index. The default is to index every word on the first page at that URL. Alternatively, the librarian can tell SAGE not to index the site at all (if the site has a lot of advertising or is itself another database like Amazon.com, it may be best to avoid a full-text search of the site). Or if the site is particularly data-rich, the librarian may tell SAGE to go one or two levels deeper than the first page: to index the first page, follow all the links on that page and index every word on each of those pages, and then again. (How deep you can go is also a function of how much disk space you have. UCSD found the default of limiting the full text search to the first page with the option to go two levels deeper if needed to be sufficient for recall and also affordable in terms of storage space.)

A full-text search is particularly helpful for sites where keywords or brief annotations are inadequate to describe the resource, or for sites that change often. Keywords can never compete with full text for recall. For example, someone looking for "mars mission in 2003" would have found nothing on the UCSD Web site with a conventional search because those words don't feature in the annotation of any of the Web sites they have cataloged. But a SAGE search for "mars mission 2003" reveals six Web pages (from NASA, the Jet Propulsion Laboratory, CNN and Discovery.com) that discuss that topic from a range of perspectives. The search engine runs weekly, updating the full text faster and more often than a librarian would typically update an annotation. This currency is particularly good for news or topical queries. And because the search is restricted to starting points initially selected by local specialists the precision for UCSD's local clientele is superior to that of a general Web search engine such as AltaVista. Additionally, SAGE includes their locally licensed content.

Even without access to a system like SAGE you can increase the relevancy of search results by providing carefully chosen annotations and keywords. These need to be as user-centered as possible, using the natural language of your customers. Annotations can be brief, and may not always be required (a list of engineering societies may suffice without notes, for instance). Consider writing evaluative annotations that go beyond the basic description (state what the resource is particularly good for). Evaluative annotations supplement evaluative filters and flags (such as "Editor's Choice") and add tremendous value to your work. Jacob's Nielsen's columns "How Users Read on the

Web,"[7] and his "Eyetracking Study of Web Readers"[8] provide helpful suggestions for page layout and journalistic style of both text and annotations. Unfortunately, because they are human-generated, keywords and annotations are relatively expensive to produce (INFOMINE estimates that their catalogers spend 25 minutes per resource to create the initial record).[9] In setting the editorial guidelines for your content authors you do need to strike a healthy balance between staff time spent and patron satisfaction gained.

With any search system it's important to periodically review the search terms being used by your customers (your search engine logs these) and to refine your vocabulary as needed to match the actual vocabulary of your patron base. It's also very important to explicitly tell people the scope of each search at the outset and to repeat the scope of the search at the top of each results page. The UCSD Libraries site (*http://libraries.ucsd.edu/*) now uses the convention of indicating the search scope with a one-line description for the link to the search page, followed by FOR and BY lines that further elucidate the action and scope of the search. Thus the link to SAGE appears like this:

SAGE
Gateway to Recommended Websites
FOR: Websites, E-Journals, databases selected for UCSD
BY: subject browse, keyword search

Search engine algorithms are improving but the holy grail of the highly intelligent single search box that will take your often-fragmented natural language query, search everything (Web sites, all the world's public access catalogs, abstracting and indexing databases, full text articles at publisher's sites, etc.), and return the single perfect result is still out of reach. As of this writing the California Digital Library's Searchlight system at *http://searchlight. cdlib.org/cgi-bin/searchlight* is a pioneering effort in this area that is still under development.

PHYSICAL LIBRARY vs. VIRTUAL LIBRARY

Another dilemma facing library Web designers is how to deal with the dichotomy of subject resources, which are independent of time or space, and the library hours and services, which are usually based upon the physical or administrative structure of the library. It may look neater to keep resources on the Web pigeon-holed into separate areas for each physical library, but using those Web sites then requires your patrons to know too much about your physical structure and doesn't serve the increasing numbers of interdisciplinary researchers (who need to use multiple subjects that may be in multiple physical libraries) very well. We need user-centered design, not library-centered design. Do what you can to liberate information from the structure of your library. Also do what you can to locate context-sensitive places to point to your services from within your subject pages, to avoid the feeling that your services are totally divorced from your content. Base your design on analysis of user's tasks and workflows. This is, of course, easier said than done.

FILTERING, ZONES AND PERSONALIZATION

Roughly two thirds of users go to Web sites seeking specific information,[10] and are eager to leave as soon as they find it.[11] But the Web is so flooded with information that people are increasingly frustrated at their inability to find the specific answers that they need, and educators are concerned that students are not critically evaluating the information that they find.

What we need are smarter Web sites that can assist people in finding the most relevant and authoritative sources. Web sites need to function more like human reference librarians by providing more evaluative guidance and directional assistance. Well-chosen subject and category divisions are a start (especially helpful are the smarter categories that reflect what people actually ask for, like spectra, materials data, cost figures, market analysis data, software, sound files, and pictures). But people also need help with where to begin and what strategy to pursue.

Search filters are one approach to directing people to the most relevant sources for their particular needs. There are many types of filters, but the best ones are realistically aligned with people's actual workflows. For undergraduates, consider implementing a "panic button" for those nights before the paper is due when the students suddenly realize that they have not even started their research and the library is closed for the night. Such a filter would select those resources appropriate to the undergraduate readership level that are available instantly in full-text from computers outside of the library. Actually, everyone appreciates an "estimated time of arrival" filter, which allows you to choose between full text sources available immediately, sources you have to go to the library during open hours to get, and those sources that will take days or weeks to arrive via interlibrary loan. Data Genie (*http://www.lib.calpoly.edu/research/data_genie.html*) from the California Polytechnic State University at San Luis Obispo provides a filter like this by asking online users how long they are willing to wait to receive the information.

Also in the category of time-related filters is the new books list, or "what's new" page listing newly added Web links. While these are helpful, consider creating a more targeted and proactive update service that would allow patrons to sign up for custom news releases from the library targeted to their research interests, or to be automatically notified by e-mail when library Web pages (or lists of subject resources) they have selected have been updated. Free services like NetMind (*http://www.netmind.com*) and EoMonitor (*http://javelink.com/cat2main.htm*) are available for people to use to track changes in Web pages. What's needed is a simple version of this function that doesn't require excessive pre-profiling (simply "click here to be notified when this page is updated" is sufficient) for library Web sites. If you could also offer to seamlessly integrate the library Web page updates with updates from your abstracting and indexing databases that's even better.

For people new to a discipline (be they undergraduates or faculty new to the field) consider a "for starters" filter (or flag) that identifies review articles or other overviews of the field. (Let's take advantage of the "material type = review" field that so many bibliographic databases provide). Other evaluative filters such as "editor's choice" or "key sites" are also increasingly necessary to distill the ocean of information down to a trickle of the most authoritative sources.

Filters depend upon the underlying metadata that you have applied to each resource. If you think in terms of the specific tasks that people want to do (and even better, if you actually perform a task analysis of your customers), it becomes easier to identify the type of metadata you will need to capture. For instance, real people often ask for reference books by color. Librarians know this is a temporary quality, as colors change with editions or re-binding, but it represents a real-world need. The New England School of Law Library provides a search interface to their reserve books that allows people to search for books by color (see *http://portia.nesl.edu/screens/well_its_red.html*). This is a wonderful application of user-centered metadata.

Academic libraries serve at least four (and often more) distinct populations: faculty, graduate students, undergraduate students, and staff. Another approach to getting people connected to the specific information that they need (and one that uses browsing rather than searching) is to create a different view or zone of your Web site specifically tailored for each user group. The Science & Engineering Library at UCSD did this in their fall 1999 Web site redesign by creating the Faculty Zone, the Grad Zone, and the Undergrad Zone. In this model you ask people to self-select the view of your site most appropriate for them. Unfortunately, research has found that users resist self-characterization, and that asking users to define their roles is not a useful strategy for guiding them through a site.[12] At UCSD the zones were fairly well used (the Undergrad Zone homepage ranked as the tenth most popular page out of over 50 library services pages, garnering between 1250 and 1705 views per quarter in 1999-2000) so this approach should not be characterized as completely useless. Nevertheless, in the fall 2000 redesign they retooled the zones from a role-based to a task-based model. The faculty and grad zones became a single

Research Zone, and the undergraduate zone became a classroom or "How Do I . . . ?" Zone.

Finally, a brief mention of customization and personalization is in order. These are two more approaches to helping people manage large amounts of information. As defined by Jacob Nielsen,[13] customization is under direct user control: the user explicitly selects between certain options (Data Genie's how long are you willing to wait filter is a form of customization). Personalization, on the other hand, is driven by the computer that tries to serve up individualized pages to the user based on some form of model of that user's needs. Amazon.com's practice of welcoming repeat visitor's by name and offering them purchase suggestions based upon their past purchasing history is an example of personalization. The *MyLibrary@NCState* project (*http://my.lib.ncsu.edu/*) from the North Carolina State University Libraries in Raleigh is an example of customization applied to a library Web site that is generating lot of well deserved positive interest, particularly since they are willing to share their source code. For more information about *MyLibrary@NCState*, see *http://my.lib.ncsu.edu/about/*.

ELECTRONIC JOURNALS

Obviously, electronic journals are a prime resource, particularly in the sciences. Currently, many institutions maintain two modes of access to electronic journals: via their Web OPAC and via the engineering library Web page. Professional cataloging of e-journals is wonderful if your institution has the staff to provide this service in a timely manner and if you can get all of the categories of e-journals that meet your quality/relevancy requirements included. Ideally, you want access to selected trial subscriptions, titles for which there is no local print equivalent as well as those with print, and selected free electronic journals if they are of high quality.

Relying exclusively on the OPAC to provide access to your e-journals insures you of the quality and authority control inherent in professional cataloging, and does a great job of integrating your e-journals with the print collection. But OPAC access alone segregates e-journals from the rest of your Web site. Also, catalog displays have to date been limited. For example, patrons who want to quickly scan a list of all of your e-journal titles in materials science (say, 150 titles or more) to see which ones start their online coverage as early as 1993 are often frustrated by catalogs that only display ten records per page and that require each record be opened individually to determine dates of coverage. For these reasons, and also to allow for the addition of real-time notes added by bibliographers which may stretch the capacity of the MARC record or the institutional authority of non-catalogers to access the OPAC (notes such as "site temporarily unavailable," or "for the article on sheep cloning, see volume 31 page 17"), many libraries have also provided a separate master electronic journal Web page as well as subject-specific e-journal pages. To code these pages by hand is tremendously time-consuming, particularly since the publisher's Web sites change more often than you would imagine and the same title may need to be on multiple subject-specific pages.

Ideally, you want to catalog electronic journals once, then use those records in multiple places both inside and outside of your OPAC. This can be achieved by starting with the OPAC record, then exporting those records directly into your Web site database or into another file or database that is then accessible both to the bibliographers (who add the real-time notes) and to the scripts that generate your Web pages. Several libraries have used a variation of this exporting method successfully. The Los Alamos National Laboratory Library has been doing this since 1997, the University of Washington at Seattle since September 1998, and the University of California at San Diego since October 2000. For the technical details of the Los Alamos approach, see "Creating Electronic Journal Web Pages from OPAC Records" at *http://www.library. ucsb.edu/istl/97-summer/article2.html.*

OTHER EDITORIAL ISSUES

There are a host of other editorial decisions that go into designing Web sites:

- What is the best vocabulary for link labels? (Avoid library jargon and user test, but that is easier said than done.)
- What scheme should you use for subject headings for browsing? Library of Congress subject headings, or LC classification for its hierarchy? Institution majors or degree program names? Intuition? How do you keep the list to a manageable size? How do you provide scope notes (both for the public and for the content authors) and "see also" notes?
- How do you select type or format headings for browsing? Should you mix formats like "software" and "data" with genres like "dictionaries" and "encyclopedias"? How many formats are too many?
- How do you achieve authority control for authors and titles, and how much control is needed?
- What page layout is most readable on the Web?
- What writing style should you use? (See Jacob's Nielsen's columns "How Users Read on the Web,"[14] and his "Eyetracking Study of Web Readers"[15] for helpful suggestions.)
- How do you assure initial and ongoing quality control?

- If done right, user testing never ends. How do you incorporate usability testing into everyone's job description so that it is mainstreamed into your workflow? How do you optimize testing so that you do enough at the right times, but don't go overboard? (See Jacob Nielsen's columns "Why You Only Need to Test with 5 Users,"[16] "Voodoo Usability,"[17] and "The Mud-Throwing Theory of Usability."[18])

Unfortunately, to do each of these topics justice is beyond the scope of this article. In addition to the references listed above, suggestions for further study can be found in the recommended reading section of this article.

THE FUTURE OF REFERENCE ON THE WEB

In "Determining Our Digital Destiny"[19] Roy Tennant wrote, "Statistics collected by the Association of Research Libraries (available at http://fisher. lib.virginia.edu/newarl/) indicate that the number of reference queries handled per professional staff member has gone down over the last two years at a number of ARL libraries. One can only conjecture as to the reason, but my money is on the Internet."

Why should librarians be concerned about in-person reference statistics going down if they have designed and built the most fantastic engineering library Web site imaginable? Because apparently rather than visiting library Web sites, our patrons are migrating to the commercial reference services springing up on the Internet, like AskJeeves (*http://www.askjeeves.com*) and Webhelp (*http://www.webhelp.com/home*). These sites have proven wildly popular. AskJeeves serves over 5 million people per month with more than 2 million questions answered per day, while Webhelp is getting 5 million hits per day.[20] If our primary clientele are not visiting the reference desks in person, and they prefer other commercial real-time Web reference sites to ours (particularly when commercial interests may not always serve them as well as we can with our unique licensed content), then we have a problem.

Webhelp is a real-time reference service that uses Web collaboration software, allowing a live person to work with you to answer your question–just like a reference librarian–although no longer free of charge. While AskJeeves has to date only provided a database of previously answered questions through which you can search (albeit with the assistance of some natural language parsing software and some relevancy ranking technology), it has recently acquired similar collaboration software so one would expect to see it go interactive in the near future.[21] Libraries that simply provide passive access to a Web site may no longer remain competitive. Should libraries retool to provide interactive human-mediated reference service via the Web? Should they do this cooperatively with other libraries (perhaps in other time zones), so that they can provide "we never close" service 24 hours a day, seven days a week? How are these cooperative systems to be built, managed, and paid for? And what happens if they become as popular as the commercial sites? Will they be overwhelmed?

Two projects aim to tackle these and other questions. The first is the 24/7 Reference Project (*http://www.247ref.org*) from the Metropolitan Cooperative Library System headquartered in Pasadena, California. The core of the 24/7 project is a set of software tools that enables librarians to provide real-time reference assistance to their patrons through the World Wide Web. The 24/7 project software can do some pretty nifty things–through "follow-me browsing" the librarian can show the patron where to go even through a complex multi-step search strategy, or the librarian can push selected Web pages to the patron's browser through "page sharing." (The patron can also push pages back to the librarian to show them exactly what they are looking at.) "Form-sharing" capabilities allow librarians to assist patrons in filling out forms to search databases or online catalogs. The librarian and patron can communicate during the session via online chat or a live phone connection, if the patron has an extra phone line available.

However, the 24/7 project is about more than just the technology. They also plan to explore methods for effective sharing of reference services among libraries, and to work out equitable compensation models for costs incurred. Furthermore, they are constructing portal-like Web sites on selected topics such as business and consumer health that will be the launch pads for the interactions between the patrons and the librarians. The Santa Monica Public Library has been using the 24/7 software to answer questions from the public since July 1, 2000.[22] As of this writing quite a few public libraries are actively participating. UCLA is offering the service to their students in a limited fashion for a few hours a day, and several other universities are participating in trials (see *http://www.247ref.org/communities. htm* for a complete list).

The other initiative to keep an eye on is a research proposal currently under consideration for funding by the National Science Foundation, entitled "Real-Time Reference for Undergraduate Students in Science, Mathematics, and Engineering" (see *http://www.library.cmu.edu/Libraries/24x7.pdf*). The principle investigators from the University of Arizona, Carnegie Mellon University, University of Hawaii, Online Computer Library Center (OCLC), Oregon State University and the University of Texas propose to design and staff a chat room 24 hours a day, seven days a week for eighteen months. Staffing of the chat room will "roll across the time zones." The questions and answers from the sessions will also populate a database that will be used as a reference tool by both the librarians and eventually the students, to supplement the real-time assistance.

CONCLUSIONS

As information professionals we need to be players in the digital library and the digital reference game; it's a natural extension of the evolution of library Web sites. To remain competitive we must become better at innovation and building prototypes (better at rapid evolutionary development), and we need to observe the behavior of real users and design smarter systems that meet their real needs. To paraphrase Roy Tennant,[23] to remain in control of our digital destiny libraries need to begin serving users what they really want (be it print or digital), whenever and wherever they want it. Welcome to the 21st Century!

RECOMMENDED READING

Arthur, Lowell Jay. *Rapid Evolutionary Development*: *Requirements, Prototyping & Software Creation*. New York: Wiley, 1991.
Baca, Murtha, ed. *Introduction to Metadata*: *Pathways to Digital Information*. Los Angeles, Calif.: Getty Information Institute, 1998.
Fleming, Jennifer, and Richard Koman. *Web Navigation*: *Designing the User Experience*. Sebastopol, CA: O'Reilly, 1998.
Nielsen, Jakob. *The Alertbox*: *Current Issues in Web Usability*. Available at *http://www.useit.com/alertbox/*.
Nielsen, Jakob. *Designing Web Usability*. Indianapolis, Ind.: New Riders, 2000.
Nielsen, Jakob, and Robert L. Mack, eds. *Usability Inspection Methods*, New York: John Wiley & Sons, Inc., 1994.

REFERENCES

1. K. Calhoun, Z. Koltay, and E. Weissman, "Library Gateway: Project Design, Teams, and Cycle Time," *Library Resources & Technical Services* 43, no. 2 (1999).

2. Murtha Baca, ed., *Introduction to Metadata*: *Pathways to Digital Information* (Los Angeles, Calif.: Getty Information Institute, 1998).

3. James Gleick, *Faster*: *The Acceleration of Just About Everything* (New York: Pantheon Books, 1999).

4. Jakob Nielsen, *Designing Web Usability* (Indianapolis, Ind.: New Riders, 2000) 218.

5. Lowell Jay Arthur, *Rapid Evolutionary Development: Requirements, Prototyping & Software Creation* (New York: Wiley, 1991).

6. Nielsen, *Designing Web Usability* 160.

7. Jakob Nielsen, *How Users Read on the Web* (October 1, 1997 [cited August 28, 2000]); available from http://www.useit.com/alertbox/9710a.html.

8. Jakob Nielsen, *Eyetracking Study of Web Readers* (May 14, 2000 [cited August 28, 2000]); available from http://www.useit.com/alertbox/20000514.html.

9. Julie Mason et al., *Infomine: Promising Directions in Virtual Library Development* (University of California Riverside, May 11, 2000 [cited August 24, 2000]); available from http://infomine.ucr.edu/pubs/VLPromisingDirections.html.

10. Richard Koman, *Helping Users Find Their Way by Making Your Site "Smelly"* (Webreview.com, May 15, 1998 [cited August 24, 2000]); available from http://webreview.com/pub/98/05/15/feature/index.html.

11. Will Schroeder, *Steering Users Isn't Easy* [web magazine] (Sun-Netscape Alliance, 1998 [cited August 24, 2000]); available from http://developer.iplanet.com/viewsource/schroeder_ui/schroeder_ui.html.

12. Ibid.

13. Jakob Nielsen, *Personalization Is over-Rated* (October 4, 1998 [cited August 28, 2000]); available from http://www.useit.com/alertbox/981004.html.

14. Nielsen, *How Users Read on the Web.*

15. Nielsen, *Eyetracking Study of Web Readers.*

16. Jakob Nielsen, *Why You Only Need to Test with 5 Users* (March 19, 2000 [cited August 28, 2000]); available from http://www.useit.com/alertbox/20000319.html.

17. Jakob Nielsen, *Voodoo Usability* (December 12, 1999 [cited August 28, 2000]); available from http://www.useit.com/alertbox/991212.html.

18. Jakob. Nielsen, *The Mud-Throwing Theory of Usability* (April 2, 2000 [cited August 28, 2000]); available from http://www.useit.com/alertbox/20000402.html.

19. Roy Tennant, "Determining Our Digital Destiny," *American Libraries* 31, no. 1 (2000).

20. Steve Coffman and Susan McGlamery, "The Librarian and Mr. Jeeves," *American Libraries* 31, no. 5 (2000).

21. Ibid.

22. McGlamery, Susan. E-mail to author, 8 November 2000.

23. Tennant, "Determining Our Digital Destiny."

INFORMATION COMPETENCIES

Model for a Web-Based
Information Literacy Course:
Design, Conversion and Experiences

Leslie J. Reynolds

SUMMARY. When converting a traditionally taught classroom course to a web-based environment, you must take a number of issues and consid-

Leslie J. Reynolds, MS, is Assistant Engineering Librarian and Assistant Professor of Library Science, Siegesmund Engineering Library, Purdue University, West Lafayette, IN 47907. She has taught information strategies to Electrical Engineering Technology students since fall 1997 and has team-taught the course online since spring 1999.

The author wishes to acknowledge the grant from the Indiana Partnership for State-wide Education and support from the Purdue University Libraries.

The author also wishes to thank the following people for their support and assistance in both teaching this course and the preparation of this article: Sheila R. Curl, Alexius Smith Macklin, and Brent Mai.

[Haworth co-indexing entry note]: "Model for a Web-Based Information Literacy Course: Design, Conversion and Experiences." Reynolds, Leslie J. Co-published simultaneously in *Science & Technology Libraries* (The Haworth Information Press, an imprint of The Haworth Press, Inc.) Vol. 19, No. 3/4, 2001, pp. 165-178; and: *Engineering Libraries: Building Collections and Delivering Services* (ed: Thomas W. Conkling, and Linda R. Musser) The Haworth Information Press, an imprint of The Haworth Press, Inc., 2001, pp. 165-178. Single or multiple copies of this article are available for a fee from The Haworth Document Delivery Service [1-800-342-9678, 9:00 a.m. - 5:00 p.m. (EST). E-mail address: getinfo@haworthpressinc.com].

165

erations into account. This paper addresses the process of design and conversion, including computer mediated communication as a form of distance education, the advantages and limitations of the technology, methods of addressing learning styles and insuring interaction with students, as well as examples of teaching an online asynchronous course. *[Article copies available for a fee from The Haworth Document Delivery Service: 1-800-342-9678. E-mail address: <getinfo@haworthpressinc.com> Website: <http://www.HaworthPress.com> © 2001 by The Haworth Press, Inc. All rights reserved.]*

KEYWORDS. Distance education, distributed education, online learning, virtual classroom, information literacy instruction, Web-based instruction, bibliographic instruction, problem based learning, networked learning environment

In the next century, a new kind of university will be in place. Most of us are already in the process of inventing it. It will be a hybrid, preserving the best of our traditions and adapting them to meet new needs. A university without walls, it will be open, accessible and flexible in ways that can barely be imagined today. In this new university, the emphasis will be on delivering instruction anywhere, anytime, and to practically anyone who seeks it.[1]

Course websites are commonplace these days. In the last five years, the academic community has seen an increasing use of the World Wide Web for instruction. New technologies offer educators opportunities to enhance student learning and expand modes of teaching students anytime, anywhere. Members of the Purdue University Libraries faculty received a statewide grant in the spring of 1998 to convert and redesign a required, one-credit asynchronous course designed to teach information strategies to undergraduate students in the School of Technology. I will address the process of design and conversion of an online course, including computer mediated communication as a form of distance education, the advantages and limitations of the technology, methods of addressing learning styles and ensuring interaction with students, as well as share some teaching experiences from an online asynchronous course.

DESIGN OF COURSE: LINEAR MODEL CONVERSION

Designing an online course requires a significant amount of time, to plan, write and create basic lessons. A common misconception is that this is where the instructor's commitment ends. This type of instruction requires not only time for initial design, but also redesign, troubleshooting and actually teaching.

It was critical that we integrate search strategies, evaluation and citation formats, as well as a way to immediately draw in the students' interest by showing them how this course relates to their real lives. If they could not see the relevance in the beginning, they might not come back to the course site to give it a second chance. The content was not the problem–it was converting the content. Our efforts at course development began with deconstructing the traditional course. Simply taking lecture notes and placing them online was unacceptable because the lessons needed to be dynamic. The initial model designed for this course was linear and traditional (see Figure 1).

However, no matter how creative the designers tried to be, the linear outline would not flow correctly in this environment. The course needed to be modular. It needed to be flexible. It needed to be designed with the idea that the student could choose a lesson, read it, and understand it, even if they selected one

FIGURE 1

Unit One: Identification and Evaluation of Resources

- ◯ Identify and distinguish among types of information found in basic scientific and technical reference resources.
- ◯ Assess source's scope, currency, format, bias and qualifications of compilers and authors.
- ◯ Understand the structure and purpose of citation formats.

Unit Two: Online Keyword and Index Searching

- ◯ Use different search methods.
- ◯ Formulate all appropriate database queries.

Unit Three: Journal Articles and Indexing Tools

- ◯ Search for topic-specific journal articles.
- ◯ Construct a citation.

Unit Four: Specialized Literature

- ◯ Introduce information sources specific to electrical engineering (e.g., patents, standards, and product literature).

Unit Five: Search Strategies

- ◯ Formulate search strategies.
- ◯ Apply search engine strategies to new information systems.

from the middle portion of the semester in a non-linear fashion. In addition, we wanted the course designed so that sections of it could be reused for other asynchronous teaching opportunities with little editing (i.e., provide a module for another professor teaching online that needed an information strategy component).

DESIGN OF THE COURSE: CIRCULAR MODEL

We wanted the modular course to bring together abstract concepts and concrete examples. We decided that the course model should mimic the real world and how professionals create information. Evaluation of the original linear course model and its reorganization brought to light Subramanyan's model for the evolution of science and technical information. It was decided to hang the course modules off the points on his circle.[2]

Following some editing and manipulation of Subramanyan's circle, a progression for the course was developed. Figure 2 shows that in this circle, the evolution of information creation flows clockwise, from its generation as a result of research and development activities through its dissemination in primary literature, its surrogation in indexing and abstracting services and its eventual integration and compaction in reviews, textbooks and encyclopedias. The consumption of information moves in a counter-clockwise fashion. We identified "evaluation" as the central theme and placed it in the center of the circle to reflect its importance and relevance to the entire information seeking process.

The circular model of the information life cycle provided the flexibility needed to teach this course online. We deconstructed and reconstructed the component parts of the original course with the central focus on evaluation: identify an information need, then evaluate it (what do you already know, what do you need to know to meet that need); identify possible places to search, then evaluate them; locate resources, then evaluate them; use the resources, then evaluate what you found. Students were asked to think of the circle as a spiral, building upon itself. As you learn new pieces of information, you use this knowledge to create something new.

The instructors attempted to show students where information seeking fit into their lives, both personally and professionally. We incorporated some components of Problem Based Learning (PBL) into the course design. PBL is a strategy for posing significant, real-world situations, while providing resources, guidance and instruction to learners as they develop content knowledge and problem solving skills.[3,4]

In this course, we used a series of historical events that build off each other so that, as the information retrieval skills are developed, they transfer from one

FIGURE 2

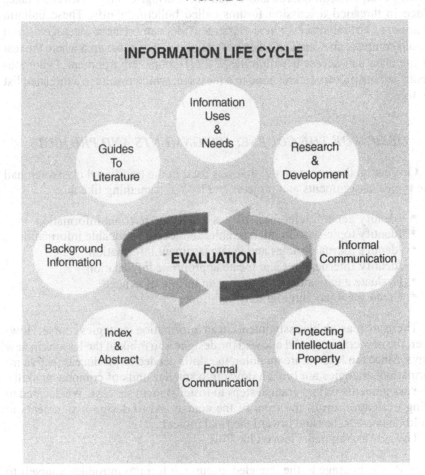

event to another. We developed two case studies: one dealing with the development of Motorola's Iridium product and service, and the other involving patent infringement at Kodak.

We used WebCT course management software to construct the asynchronous classroom environment. The designers selected this software because Purdue University had already purchased a site license and the students would likely be familiar with the interface from other courses. In addition, with the WebCT software, we could more easily collaborate with faculty and incorporate library modules into their courses. WebCT provides the instructor with management tools for grading and tracking student interaction, as well as a variety of ways to conduct computer mediated communication including private e-mail, chat, bulletin boards and document sharing. Course discussion takes place in threaded discussion forums called bulletin boards. These bulletin boards are virtual spaces where class members can create a message as you would compose an e-mail and then post it online into a topic area where the rest of the class can access it within the WebCT course environment. Other students and instructors can respond to a message, which results in a threaded list by topic.

DESIGN OF THE COURSE: ASSIGNMENTS AND PROJECTS

Originally, the linear model that was used in the traditional classroom had the typical assignments and projects and looked something like this:

- Identify general reference sources and locate relevant information
- Identify topic-specific monographs and locate applicable information
- Identify journal articles and locate pertinent information
- Identify Internet sites and locate appropriate information
- Evaluate a case study identifying an information need
- Create a subject bibliography/pathfinder

These are reasonable assignments in an information strategies course. However, the new course model allowed the designers to think of the lessons in new ways. Since the lessons are modular, the skills needed to produce effective information strategies are broken down into smaller units of component skills. The assignments reflect gradual steps to foster student success. We decided to bring evaluation up to the front of the course and to design assignments to build on one another and toward the final project.

The new assignments looked like this:

- Post a message to the threaded discussion form to introduce yourself to your classmates (questions are provided and a model for the introduction is provided by the instructor's own intro message)
- Evaluate a strange or misleading website with your team or evaluate and compare two opposing websites; take a self test
- Identify features of a search engine and journal database; take a self test
- Evaluate general and sci-tech encyclopedia articles using criteria from assignment one; take a self test
- Coordinate with team to select research question and form hypothesis for following assignments and final project
- Identify keywords, phrases and related disciplines to a given topic
- Identify, locate and evaluate pertinent information in monographs and other forms of background information

- Identify, locate and evaluate pertinent information in journal articles and other forms of current information
- Coordinate with team to select the best information for presentation as final project (simulate information need in a real-life work situation)

Course discussions, which revolve around lessons and assignments, include instructors asking open ended questions to challenge the students to think about the information provided and use the information to solve problems, create new knowledge and apply that knowledge to the assignment at hand.

DESIGN OF THE COURSE: BUILDING THE LEARNING ENVIRONMENT

In addition to using WebCT courseware, the designers embraced a new online learning model. The Networked Learning Environment (NLE) is a paradigm for teaching information retrieval, evaluation, organization, interaction and communication by involving active learning facilitated with simulated experiences. It uses cognitive modeling whereby the student builds upon what he already knows and engages in cooperative and collaborative problem solving that develop critical thinking skills.[5]

The center of the NLE model is the student–the individual learner. Four partners surrounding the learner share in the creation, interpretation, dissemination, and management of information. These partners are the human facilitators (instructors), peer mentors, computer technology and the learner's background and previous experiences that shape and create personal reality. For the exchange of information and ideas to be productive, there must be collaboration among all the partners.

Since distance learning can make students feel isolated, working with a group and sharing experiences can provide the social framework needed to remain involved in the course. The individual learner in the NLE is dependent upon the others to assist in problem-solving and share resources for data collection, organization, and evaluation. There is a built-in expectation for active learning experiences and open communication to take place in this environment. There is a continuous movement throughout the learning process as students create, evaluate, and revise information, as well as develop critical thinking skills necessary for promoting effective information-seeking behaviors. Simulating real world experiences within the NLE encourages the learner to evaluate the information they retrieve.

Students must be self-motivated, independent and be able to communicate effectively in writing, as we require students to post messages of substance at least once a week. It is easy for students to postpone attendance and then fall behind, creating the likelihood they will feel the pressure from information overload (i.e., bulletin board messages) and decide not to come back to class. We post homework hints and tricks and ask probing questions to facilitate the discussion. The bulletin board communications become the most interesting part of the course–for the instructors as well as the students.

In some ways, there is enhanced access to instructors and other students; class members can draw on group knowledge, experience and support. Instructors are in the course website several times a day, five days a week, sometimes on evenings and weekends. Students can ask the instructors questions while they are interacting with the lesson material and get a relatively quick response. Students cannot disguise poor performance in the online classroom since they must read and communicate their understanding regularly through assignments and interactions with students and instructors via chat and the bulletin boards. Active participation is required in an asynchronous course; one must make a comment to be seen as fully "present." This type of interaction and peer discourse encourages collaborative problem solving.

Instructors must monitor the student-to-student interactions to correct misconceptions as quickly as possible because of the emphasis on group support and collaboration. Students working in teams sometimes run the risk of falling victim to "group think" and then take the path of least resistance on an assignment instead of thinking critically about the problem presented. For example, one team was selecting a team topic for their final project. Their chat room discussion deteriorated into the idea that what they really needed was a good title. What they turned in was catchy, but did not meet the requirements of forming a research question. While this kind of "group think" can be problematic, monitoring and facilitating team discussions can redirect a team that has gone down the wrong path.

Each team has its own separate discussion space and an area for placing and working on shared files. Stored communications that remove the pressure of instant response facilitate reflection and thoughtful comments. Student teams are encouraged to use the chat rooms, and some of them do. The chat rooms allow students to interact with each other or an instructor in real time. When they use the chat rooms, we ask the students to select one of the four rooms that log conversations. We can capture their discussions and put a copy of the log in the team's shared file space so they can review it later. The chat room logs provide instructors an opportunity to see how the team is working together. Several teams, knowing that the instructors read the chat logs, include comments to us within their discussions.

BENEFITS AND CHALLENGES OF THE TECHNOLOGY

Learning and teaching in the online environment requires that all participants be comfortable with the technology. During a face-to-face student orientation at the beginning of the course, instructors demonstrate how to move within the course website. This is usually the first time most of the students have taken an online course and therefore the first time they have seen the WebCT interface. Students are also taught (or reminded) how to cut and paste from one document to another, how to open a frame in a new window and how to post a message to a bulletin board.

Taking an online course necessitates computer access and knowledge. Technical difficulties are probable–e.g., students have had difficulty with the network, their hardware "computer melt-downs," software and file transfer. Students bring different levels of technical expertise into the virtual classroom. Many of the students felt much of the orientation was very basic and easy to understand. However, students who did not attend the orientation had difficulty navigating the course website, had a hard time communicating with other students online and frequently missed the benefits of course participation through discussion. These students were more likely to become disengaged or drop the course.[6]

Following is a real-life example of a student who missed orientation. The student told me he was having trouble getting his team to respond to him. The team claimed that they were contacting him via e-mail and he was not responding. When the student came in again, I asked if he had recently checked his course e-mail. He said not today.

> I said, "Let's check your e-mail."
> He said, "I can get to AOL from here?"
> I responded, "Does your group know your AOL e-mail address?"
> He said, "Doesn't it just 'go there?' "
> This student missed the orientation and was inexperienced when he thought the computer automatically "knew" where to send his mail.

BENEFITS AND CHALLENGES
OF THE ASYNCHRONOUS INSTRUCTION

Since we deliver the course over the web, the students in remote locations are able to complete this course, which is required to earn an Electrical Engineering Technology degree. They interact with the subject matter, the instructor and other students when it is convenient for them. This is especially useful for non-traditional students who find it inconvenient, if not impossible, to attend classes offered during normal class times. Employed adult learners in particular, desire asynchronous instruction opportunities because of the added flexibility it provides in accommodating their studies to their business and family responsibilities.[7]

However, both traditional and non-traditional students are not yet used to working in an asynchronous online environment. In order to be successful in an online course, they must be sufficiently self-disciplined to attend class. That sounds simple, but traditional students have spent the last twelve or so years of their lives attending classes when and where someone told them to go. For some students, the freedom that this type of course allows provides challenges to their time management skills. One student who dropped the course reported that without a particular seat to occupy at a predetermined time, she could not work it into her daily schedule.

Development of the student-to-student and student-to-facilitator relationships is essential for asynchronous instruction to be successful. Online rapport is built by sharing ideas; participants get to know each other through e-mail, threaded discussions, etc. Part of learning is not just getting the facts, it is the sharing of experiences and stories that relate course content to real-life situations. This makes the class meaningful and engaging. Students have an opportunity to think before they speak, whether it is five minutes or five hours. They compose what they want to say after they have time to reflect. Students who would not normally talk in class (perhaps because they have not had time to process the discussion) can confidently and actively contribute in a non-threatening and variably paced environment.

The instructors and the students define the concept of asynchronous learning differently. As an instructor, I find it relatively easy to be at the computer, or at least have it turned on, eight hours a day. Keeping up with discussions and e-mail is not an enormous burden. We found it easy to go into the course area for a few minutes several times a day. Students get to the computer when they can. For students who own their own computer, that usually means the wee hours of the morning. Students who do not own their own computers had to rely on computer labs and computer availability within those labs. We found that the most successful students came into the virtual classroom for a few minutes every day.

Teaching in a traditional physical classroom, the instructor can tell through verbal clues and body language if a student is paying attention and understanding the lesson content. Online, this observation is much more difficult. The students must feel comfortable asking questions and openly discussing content. Students who do not understand a lesson and do not "speak up" in a web-based course may have a difficult time completing assignments successfully. Whereas facial expressions help to identify a student having trouble, teaching online means the instructor may not know that a student is floundering until he turns in an assignment and does poorly or fails to turn in any assignment.

Writing assignments with concrete instructions has been a challenge for instructors.[8] In a traditional classroom, you hand out the assignment, discuss it, ask if there are any questions. You can tell by students' reactions and body language whether or not they fully understand it. In the virtual environment, that type of feedback does not exist.

After the first semester the course was taught online, we went back over our assignments, removed jargon and made directions as concrete as possible. This refinement, in addition to soliciting student feedback during the assignment, resulted in better work from the students, and we continue to review and refine assignments to update and improve them.

The problem is not always how the assignment is written. One of the early assignments asks the student to create a table to compare database search features. Last semester, two students did not understand the assignment and turned in some very strange work–a significant part of their lack of understanding was in trying to complete the assignment without reading the accompanying lesson. The student tracking module of the WebCT courseware enabled us to tell that the students had not visited the lesson pages.

This is an example of how online instruction can offer benefits to an instructor that are not available in traditional classroom instruction. The student tracking feature on the course website allowed us to see if a particular student had accessed the appropriate lesson for the assignment. When a student has difficulty with an assignment, we can check to see if the student read or at least viewed the supporting material. The students are told during the orientation session that this type of tracking is possible, and they get to see an example of it. We did not want the students to feel as though we were spying on them, but it was important to us that the students fully understood what kind and how much information about them and their activities in class we gathered.

DESIGN OF THE COURSE: ADDRESSING LEARNING STYLES AND ENSURING INTERACTION

Students have different learning style preferences in the ways they process information. Some students focus on facts and data; others are comfortable with theories and abstract ideas. There are students who respond strongly to visual forms of information; others get more from verbal forms of information such as written and spoken explanations. Some prefer to learn actively and interactively; others function more introspectively and individually.[9]

Regardless of the learning environment in which you teach, it is essential that the students are vested in "coming" to class. Students do not walk into a classroom and literally sit in a seat, so instructors must find a variety of means to anticipate and meet their learning needs. Interaction among students, instructors and the learning environment is important for successful distance learning. One way to insure interaction and address different learning needs is to incorporate synchronous opportunities for discussion of issues among the students or with the instructor. This course begins with a face-to-face orientation and incorporates real time chat rooms for the purpose of office hours. Recognizing the need of some students to have real-time engagement, we use online chat rooms for group projects and additional student-to-student and student-to-instructor communication.

Other ways to encourage interaction in the digital environment are similar to those used in a regular classroom situation, for example, asking a question of a particular student or asking a student to lead a discussion. Cooperative learning is also possible in cyberspace; the students can successfully work in virtual groups to complete projects and assignments.

Asynchronous instruction requires planning lessons completely, including writing lesson texts (anticipating where misunderstandings may occur), creating visual examples and images that support the text, generating ideas and questions for discussion and designing self-tests and/or assignments to reinforce lessons. In a face-to-face classroom situation, it is possible to jot down a couple of notes about the points that you want to cover, go into the classroom and talk off-the-cuff about your goals for the day–this is an impossible situation in the asynchronous classroom.

Whether teaching in a traditional classroom or in an online course, the learning objectives which successful students achieve include:

1. Recognizing the importance of building research methodologies in relationship to successful information retrieval and analysis.
2. Practicing the key concepts of information retrieval, e.g., evaluation methods for information, characteristics of electronic databases, distinctions among types of reference tools.
3. Demonstrating ability in finding information in a variety of resources.
4. Selecting, analyzing and synthesizing relevant information.
5. Organizing and presenting this information effectively.
6. Valuing the role information plays in real world problem solving, including academic success.

These objectives are the measurement of success in student learning and academic performance. The outcomes expected as a result of instruction are identified for the learner in the syllabus, and reinforced throughout the course in planned instructional materials, e-mails, and chat room discussions. At the very core of our course is the notion of the learner-centered classroom. The students in this course have very diverse backgrounds and experiences, which greatly influence their learning. In planning the lessons' content, each new activity builds off the knowledge learned in the previous activity, providing more opportunities for success as each student progresses through the various levels of basic knowledge, comprehension and synthesis.[10]

TEACHING A WEB-BASED COURSE:
THE REST OF THE STORY

In the virtual classroom, instructional challenges are magnified in course development, design and delivery. Among these challenges are learning course management software, initiating and maintaining student motivation, enabling group learning and communication, and ensuring clarity of instructional materials and assignments.

For the students, we have to remember that they have a normal course load–they attend classes in traditional classrooms and labs or they have full-time jobs and families–they access our course from computers in the campus computer labs, from home and from the computers in the libraries. Students come into the course for a discrete amount of time for a specific purpose–read a lesson, take a self-test, post to a discussion, send e-mail, or submit an assignment. They may be back the next day or it may be several days.

Delivering a course over the Internet provides students with flexibility–the materials are available at any time from anywhere as long as the student has access to a computer with an Internet connection. With this flexibility comes responsibility for completing course requirements in a timely manner. We base a part of the student's participation grade on not only posting responses to questions, but also reading what others have posted.

The students who are most successful in this course login each day and spend a few minutes (5-10 minutes). Coming to the course website regularly is analogous to coming to a traditional class. Students who keep up with readings and assignments and participate with their group are more likely to successfully complete the course. Preliminary analysis of the student tracking module of WebCT shows a strong correlation between frequency of visits and grade earned. In addition to online interaction, some students need the reassurance of seeing an instructor and discussing course assignments and lessons to reinforce that they understand the material.

During a typical day, I come into the office, check my e-mail and log into the course website to catch up on the discussion the students continued while I was at home and in bed. I will go in and out of the course website all day long to facilitate discussions and answer questions. My personal experience with teaching online means that students actually come to our offices or telephone for clarifications, questions and to just talk. When this course is taught in the traditional classroom, if one student comes to office hours all semester, it is surprising.

CONCLUSION

Online asynchronous instruction has both its challenges and rewards. Online courses require a significant amount of time to plan lessons, create visual examples, generate discussion questions, design self-tests and assignments, as well as teach. Students committed to learning who are self-motivated, independent and able to communicate effectively in writing are most likely to succeed in the virtual classroom. The lack of face-to-face interaction, however, can frustrate students, so having the students work collaboratively in teams and keeping course discussions informal and friendly are important to maintain involvement and build student-to-student and student-to-faculty rapport.

BIBLIOGRAPHY

1. "Returning to Our Roots: The Student Experience," *Report of the Kellogg Commission on the Future of State and Land-Grant Universities*, National Association of State Universities and Land-Grant Colleges, April 1997.

2. Subramanyam, Krishna. *Scientific and Technical Information Resources* (New York: Marcel Dekker, 1981) 5.

3. Mayo, P., Donnelly, M. B., Nash, P. P., & Schwartz, R. W. "Student Perceptions of Tutor Effectiveness in Problem Based Surgery Clerkship." *Teaching and Learning in Medicine* 5: 4 (1993): 227-233.

4. LaLopa, Joseph. "Problem Based Learning." Paper presented at the Workshop on Problem Based Learning. West Lafayette, IN, January 1999. (unpublished).

5. Smith, Alexius E. and Leslie J. Reynolds. "Reaching the Techno-stressed: Using Networked Learning Environments to Break Barriers." in *Computers in Libraries Proceedings*, Arlington, VA, March 1999: 136-142.

6. Reynolds, Leslie, Sheila R. Curl, Brent Mai and Alexius E. Smith. "Righting the Wrongs: Mistakes Made in the Virtual Classroom." in *Proceedings of the American Society Engineering Education Annual Conference*, St. Louis, MO, June 19-21, 2000. Available: *http://www.asee.org/conferences/search/20528.pdf*.

7. Swain, Phillip H. "Report by Director of Distance Learning." *Purdue University Faculty Senate Minutes* (20 October 1997), Appendix C: 10.

8. Hara, Noriko and Rob Kling. "Students' Distress with a Web-based Distance Education Course." *Center for Social Informatics Working Paper* 01-01-B1 (March 30, 2000). Available: *http://www.slis.indiana.edu/CSI/wp00-01.html*.

9. Felder, Richard M. "Matters of Style." *ASEE Prism* 6 (December 1996): 18-23.

10. Bloom, Benjamin S. ed., *Taxonomy of Educational Objectives: The Classification of Educational Goals* (New York: David McKay Co. 1956).

Industry Expectations
of the New Engineer

Ronald J. Rodrigues

SUMMARY. Commercial engineering databases as well as digital documents, e.g., journals, patents, standards, etc., are experiencing a period of rapid growth. The last thing an engineer can afford to do is spend time sifting through piles of information without useful results to show for the effort. In the corporate world, new engineers are often called upon to perform a variety of research related tasks, e.g., report to their group on newly emerging technologies, or help in patent prior art research. Engineers with solid library research skills will generally produce more thorough reports than those without. The ideal time for the engineer to develop his or her information gathering and management skills is not when entering the corporate world, rather, it is during the engineering education where engineering library resources in staff and collections are virtually always superior to that of the corporate world where library service may be limited or non-existent. *[Article copies available for a fee from The Haworth Document Delivery Service: 1-800-342-9678. E-mail address: <getinfo@haworthpressinc.com> Website: <http://www.HaworthPress. com> © 2001 by The Haworth Press, Inc. All rights reserved.]*

KEYWORDS. Engineers, undergraduate education, information gathering, library skills

Ronald J. Rodrigues, BA, MLS, is Information Research Consultant, Agilent Technologies Laboratory, Palo Alto, CA 94304-1392.

[Haworth co-indexing entry note]: "Industry Expectations of the New Engineer." Rodrigues, Ronald J. Co-published simultaneously in *Science & Technology Libraries* (The Haworth Information Press, an imprint of The Haworth Press, Inc.) Vol. 19, No. 3/4, 2001, pp. 179-188; and: *Engineering Libraries: Building Collections and Delivering Services* (ed: Thomas W. Conkling, and Linda R. Musser) The Haworth Information Press, an imprint of The Haworth Press, Inc., 2001, pp. 179-188. Single or multiple copies of this article are available for a fee from The Haworth Document Delivery Service [1-800-342-9678, 9:00 a.m. - 5:00 p.m. (EST). E-mail address: getinfo@haworthpressinc.com].

INTRODUCTION

During an engineer's undergraduate education, for expediency he or she is often directed to specific information sources to resolve well-structured problems.[1] Corporate engineering librarians often refer to this practice as being "text-book taught." In the later undergraduate years, the aspiring engineer is increasingly exposed to situations where the problems to be resolved or the information needed to resolve them is not so clear. In this process, the nascent engineer is being introduced gradually to unstructured, real-world problem-solving and must develop strategies and tools for solving problems efficiently.

The later years of an engineering undergraduate education and the graduate years are the best time to develop a strategy and methodology for effectively identifying information needs, determining what's available, and collecting it efficiently as key steps in defining and solving technical and other problems that a working engineer can expect to encounter. These years are best because an engineering student generally has an excellent support structure available in the teaching and library staff and access to libraries, which are usually unrivaled by those to be found in any but the biggest private companies, as well as the time to use them. Problem solving by engineers, be it in research or in providing technical sales support to customers, needs to be guided by an underlying purpose. In private industry, improving financial return is usually the prime purpose. A key part of problem identification and solution in industrial situations is identifying and gathering appropriate information, often a time consuming and costly activity.

Most professional engineers have received some training in searching for information, and often have a familiarity with a range of journals and basic texts in their profession. New engineers probably belong to a professional society or two and attend local-chapter meetings regularly, and may know some knowledgeable information vendors. They usually have technical peers in their own organization and at least a small network of professional friends and acquaintances in other organizations with whom to discuss technical matters. Increasing numbers also have access to the Internet and the "free" data resources to be found there, although the Internet and the World Wide Web can entice engineers into many unprofitable hours of net surfing if they don't understand the basics behind searching for information.

Newly hired engineers often receive a "rude awakening" when faced for the first time with the level of library services available to them. Those who join Fortune 500 companies will more than likely have access to a corporate library with various levels of expertise in engineering, physical science, business, marketing, and law. However, smaller companies, start-ups and even some large companies such as Cisco Systems and Apple Computer do not offer much in the way of library services. Even at companies with a library, engineers will find that many of the bibliographic problem solving tools they used in college are absent.

Corporate libraries are referred to as "special libraries," defined as any library that caters to specific groups of people with special needs. The corporate R&D library caters to the technical community and is often extremely limited in physical space. In corporate R&D libraries, time and space directly equate to money. Academic and public libraries are generally thought of as "just in case" libraries because they acquire materials in anticipation of the needs of the users. Corporate R&D libraries, on the other hand, are "just in time," in that they deliver information to researchers at the time they need it. Many companies with libraries offer some or all of their employees database access, usually on an intranet website.[2] The ability to access these databases directly from the "computer desktop," is the trend, due partly to the great demands placed on library staff and partly to the need to make information available throughout a company that may have several locations but only one library.

Engineers in companies without library services have the double challenge of trying to find the answers by any means available to them while always being mindful of keeping the data gathering costs in hand. A recent report[3] revealed that end users (defined as those who search for information themselves, rather than using the library staff) were spending an average of 20.6 hours a week obtaining, reviewing and analyzing information. The report went on to say, ". . . half that time is spent gathering and looking for the information. The remaining half is spent reviewing and analyzing it."

Obtaining information from colleagues helps, even though it may be limited and certainly is not thorough. Nearby universities with engineering schools may also be of limited use. Engineers in these situations may take advantage of various commercial online services, e.g., Dialog's DialogSelect, or Engineering Information's Engineering Village. Both services are aimed at the engineer who values information, and has no other option available. Each service also contains a suite of information products, e.g., access to abstracts of technical papers, patents, standards and specifications, and a collection of useful related web sites. It may be cost effective for engineers without libraries to contract for services from independent information brokers, many of which specialize in science and engineering literature searching and document delivery.

EXPECTATIONS

Corporations expect their newly hired engineers or scientists to be able to "hit the ground running." Commercial laboratories often issue laboratory notebooks to new engineers on the first day of work. Laboratory notebooks are legally admissible records of the engineer's research which are used to document or establish the date of conception of an invention, or to document the reduction to practice of an invention. Possessing fundamental bibliographic skills in both technical and business research serves the engineer to more efficiently defend the company's intellectual property.

Making project deadlines contributes to a company's bottom line, i.e., profitability. By gaining expertise in using online databases, the new engineer can supply needed new ideas and information for engineers active in developing new products or upgrading existing ones, improving processes, doing root cause analysis of equipment-failure problems, integrating different kinds of technology into their products and processes, looking for new suppliers or components, investigating new uses for their products, and checking on their competitors. In order to efficiently meet these expectations, engineers would greatly benefit by knowing how to quickly locate references that meet their needs. They should be aware of all options available to them, whether funded by their company or available through a university or information broker. Far too many engineers are completely unaware of the information resources that their companies have to offer.

Information needs in a corporate environment cover a wide range of topics. In addition to finding specific data such as ranges of measurement or physical properties, engineers may encounter any of the following areas for research:

- Find out if an experiment has already been done, to avoid needless repetition.
- Find people who are the recognized experts in their field, as demonstrated by the frequency with which they are cited by other authors.
- Locate consultants or organizations that can answer complex questions.
- Find trends in venture capital spending.
- Find contract and other revenue opportunities from various governmental sources.
- Find licensable technologies.
- Identify research frontiers where efforts are most likely to lead to technical and financial success.
- Identify others who have been working on a technology of interest. It is becoming increasingly important to monitor competitors and their marketing strategies and product offerings and developments. Many large companies ask their engineers to be mindful of emerging startups for possible acquisition.
- Locate and create patents and other intellectual property. This is an extremely important activity in today's corporations. The number of patent applications continues to increase dramatically.

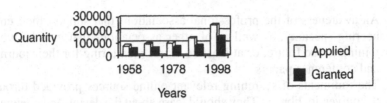

The growth of published U.S. Patents 1958-1998

Source: Science and Technology Almanac. Phoenix, Arizona. 1999. Oryx Press

- Stay in touch with changing technology and product life cycles from infancy to success by using tools that provide bibliometric analysis. It is possible to learn a great deal about a company by analyzing its published papers and/or patent portfolio.

The infancy of technological products is often marked by a peak in patents and technical reports, its commercial emergence by a peak in conference papers, and its maturity by a peak in journal articles, all measurable in many commercial databases. Online databases provide the trained engineer with cost-effective powerful search capabilities and features such as Boolean searching, proximity operators and relevance ranking. "Alerting" and "table of content" services provide the engineer with new material of specific interest and may be sent directly to his or her e-mail, dramatically cutting the time spent on reviewing hardcopy journals and reports.

THE ROAD TO ENGINEERING LIBRARY LITERACY

Engineering students, in preparation for efficiently managing information during their careers, should be departing the university for industry with more than just an accumulation of textbooks and course notes to take with them. It would be of great value for them to also possess the following proficiencies:

1. A basic knowledge of how a typical engineering library is organized, as well as a familiarity with general and engineering-specific reference books.
2. A working knowledge of the nature and usefulness of a wide range of technical journals relevant to the field. However, no matter how wide this range is, the wise engineer will realize that, with approximately 40,000 journal titles in current circulation and the number still increasing, it is not possible to be familiar with all those that may be needed.

3. An awareness of the professional associations that support their engineering specialty, as well as of other associations that may be worth joining in the future, or at least be worth monitoring for their journals and conference papers.
4. The rudiments of searching relevant online sources provided through the university library. They should learn about the design and content of these databases, not overlooking non-technical ones that provide information about competitors, suppliers, products, management techniques, and other matters that engineers and engineering managers often become involved in. A recognition of the pitfalls involved in searching the "open internet" versus the commercial online databases–the World Wide Web is far from bringing the world's information sources together into one easily searchable pool. The Internet adds many new disconnected pools of information, each with its owns rules of access and structure. Engineers should always be conscious of the time it takes to complete tasks on the Internet, and repeat the tasks using other information resources to become familiar with the advantages and disadvantages of different approaches. It is important to cross-check information from the Internet against other basic references.

TIMELINESS

This is often a major success factor, particularly in developing new products. For example, McKinsey & Co. Research[4] indicated that high-tech products delivered six months late earn one-third less profit over five years, whereas products delivered on time but 50% over budget diminish profit by only 4%. Timeliness counts too in countering competitors' actions. Sailing's premier trophy, the America's Cup, successfully defended by the United States since its first win in 1851, was lost for the first time in 1983 when the Australia II won thanks largely to the revolutionary design of its winged keel. The design was kept secret throughout the race series by its designer, Ben Lexcen. If the American defenders had looked at patent records (available in several online databases), they would have found that Lexcen had patented his design days ahead of the start of the races. Perhaps the outcome of the closely fought 4 to 3 race series would have been different if the threat had been better understood in time.

VALUE AND COST

The value of a corporate library and the desktop services it may offer cannot be overstated.[5] If at all possible, employees who are new to the company should seek out and work with librarians on complex searches and/or current awareness strategies. The corporate library also provides users with the best price for content. Engineers without the guidance of a corporate library are generally unaware of the available options for procuring documents.

In gathering information, an engineer should continually strive to balance the costs of obtaining the information against the value of the information, in order to maximize financial return.

Engineers should be trained to recognize those factors that affect this balance, and should learn to appreciate the fact that acquiring information has both direct costs, such as cash payments to external information providers, and indirect costs, such as the use of the engineer's time in conducting a search. It is also important that engineers recognize that the money saved by not doing a search may be insignificant compared with the cost of decisions based on incomplete information.

ACCESSIBILITY OF INFORMATION SOURCES

While many technical problems are resolved adequately with information at hand, others would be better dealt with by obtaining information physically remote to engineers, either outside the company or even in distant parts of their own company. For example, many people assume that product innovations are typically developed by product manufacturers, and they would look no further than their own company and key competitors for opportunities or competitive threats. However, Von Hippel's[6] research has demonstrated that both suppliers of innovation-related components and users of products can be responsible for a majority of innovations, a fact that suggests information on external technical developments is essential to stay competitive.

Nevertheless many people infrequently seek information at a distance. It has also been shown[7] that people just thirty feet apart have a probability of communicating during a week only one-third that of people closer together, and the problem worsens as the distance increases. Online services are effective at making valuable remote information accessible.

RELIABILITY AND CONTENT OF INFORMATION SOURCES

Time and effort are required to match databases to specific information needs (i.e., which databases should be consulted for this specific question) and then to learn how to use any specific database. For this effort to be worthwhile, a database user must have confidence that a specific information source will continue to be updated, and will be available in the future.

To increase the chance of obtaining the right information, the engineer should consider the breadth of content in an information source, with reference to the type of information he or she needs. All of the needed pieces are unlikely to be in just a few references. Part of the engineer's job is to intelligently put together a "mosaic of information" from what's available, or failing that, to pull the first thread of information that leads in the direction of the answers he or she needs. For example, a company that had been handstringing pearls was looking for a supplier of a machine to automate this job. An online search didn't turn up one, but it did lead to an association of fine jewelry manufacturers that supplied the answer.

Reliance on external information sources requires judgment about the quality of the information. Engineers often rely on a professional society or other authority to supplement their own judgment. Many online databases are provided by such authorities.

SEARCHABILITY OF CONTENT

One key goal of searching a database is "precision," i.e., to retrieve just those records that answer the current question, and miss as few of them as possible. Another is "recall," i.e., to gather a comprehensive set of records on a given topic, while also finding some marginally relevant material. There tends to be a trade-off implicit between these goals,[8] and the term "searchability" implies a small degree of trade-off between precision and recall. In addition to the skill of the searcher, searchability depends on the design of both the search process and the database. Experience has shown that searchability is greatest if Boolean combinations of key words and phrases are sought in specific fields within records that are divided intelligently into numerous logical fields. Being able to specify the proximity of the words or phrases within the fields improves the searchability further.

LEVEL OF EFFORT

Busy engineers are under pressure to finish assignments. Acquiring external information takes time (as much as one-quarter of an engineer's time can be spent acquiring and analyzing information),[9] and can cause costly delays. If they can propose a feasible solution to a problem without looking further, they're inclined to do so, thereby risking missing the better alternatives that additional information may have revealed. Information savvy engineers can streamline the process of finding outside information, eliminating this time pressure.

KNOWING WHERE TO LOOK

Commercial online information services, such as Dialog, Engineering Village, ISI Web of Science, are available throughout much of the world and they provide value by meeting the eight criteria described above better than other information sources in many situations. They are the basic building blocks of the "just in time" library.

Effective searching of the many databases provided by these traditional services usually requires collaboration between the person who will use the data and is familiar with the subject matter being searched, and a person who knows the range of databases being searched, their structure, and the command language used to search them. Engineers who do a lot of online searching may choose to combine the end-user and information professional roles and work alone,[10] but a partnership between an engineer and an information professional, such as a corporate librarian, is often most effective.[11] In addition to being able to conduct online searches proficiently, information professionals will know which questions are best answered by an online search and which by other means. However, the reality is that far too many engineers do not take advantage of the information resources available to them. Often the fundamentals of information literacy are woefully inadequate.

CONCLUSION

Engineering literature is growing exponentially and beginning to move more quickly towards a digital future. This is evidenced by the growth and development of commercial online services and projects like CrossRef, a publishers' initiative that provides engineers with the ability to easily move from a reference in an online journal paper to the electronic version of the cited journal article. Commercial online engineering databases are growing as well. Between 1990 and 2000 two of the major commercial engineering databases, i.e., INSPEC and Compendex, grew from a total of 6,043,413 records (3,503,563 and 2,539,850, respectively), to 11,376,871 records. They added 2,144,435 and 3,200,909 records respectively over that 10 year period. By the 3rd quarter of the year 2000, each added 220,665 and 157,430 records, respectively. Dialog alone boasted a total of 77,852,423 records for the engineering and physical science disciplines.

In today's complex global marketplace, being able to navigate the ever growing mountain of technical and business information is becoming more of a challenge. Engineers that use information well, not only have a competitive advantage over those that don't, they avoid the cost of being misinformed which can be devastating.

The transition from student to working engineer can be less traumatic when the principles above are fully appreciated. Demonstrated bibliographic research competency will help shore up the engineers, confidence that, in not only having the skills to cut time obtaining information, they will have more time to apply the knowledge when and where it counts.

BIBLIOGRAPHY

1. Leckie, G.J. and A. Fullerton. "Information literacy in science and engineering undergraduate education: Faculty attitudes and pedagogical practices." *College & Research Libraries* 60 (January 1999): 9-29.

2. Doran, K. "Delivering value-added Internet services in a corporate library." In *Change as Opportunity: Proceedings of the 88th Annual Conference of the SLA*, 139-144. Washington, DC: Special Libraries Association, 1997.

3. Anonymous. "Study on technical staff time." *EBriefs*, Outsell Inc. Burlingame, Calif., April 20, 2000.

4. Weizer, Norman et al. *The Arthur D. Little Forecast on Information Technology and Productivity*. New York: John Wiley, 1991.

5. Portugal, Franklin H. *Valuating information intangibles: measuring the bottom line contribution of librarians and information professionals*. Washington, DC: Special Libraries Association, 2000.

6. Von Hippel, Eric A. *The sources of innovation*. New York: Oxford University Press, 1995.

7. Urban, Glenn and John R. Hauser. *Design and marketing of new products*. Englewood Cliffs, N.J.: Prentice Hall, 1993. 2nd ed.

8. Zimmermann, Roy and David Erlandson. "Power searching at Teltech: gaining power through customer focus." *Database Searcher* 7(September 1991): 32-37.

9. Cloyes, K. "Corporate value of library services." *Special Libraries* 82(1991): 206-213.

10. Goldmann, Nahum. *Online Information Hunting*. Blue Ridge Summit, PA: Windcrest/McGraw-Hill, 1992.

11. Saracevic, Tefko and Paul B. Kantor. "Studying the Value of Library and Information Services in Corporate Environments: Progress Report." *Proceedings of the ASIS Annual Meeting* 35(1998): 411-425.

12. Allstetter, William, ed. *Science and Technology Almanac*. Phoenix, AZ: Oryx Press, 1999.

MANAGEMENT

The Implementation
of Information Technology
in the Corporate Engineering Library

Robert Schwarzwalder

SUMMARY. The corporate engineering library has been shaped by a variety of recent business and technology trends. The increasing globalization and focus on efficiency in the corporate world has resulted in the consolidation or organization of corporate libraries and the outsourcing of non-essential functions. The rapid adoption of Intranets and other information technologies in corporations has created opportunities for libraries to develop new services, expand their scope of influence, and serve non-traditional customers across the enterprise. This refocusing of the corporate engineering library based upon corporate objectives and

Robert Schwarzwalder, BS, MLS, PhD, is Manager, Library Systems and Information Research, Ford Motor Company, Dearborn, MI 48121 (E-mail: rschwar3@ford.com).

The views expressed in this article are those of the author and do not reflect the position of Ford Motor Company.

[Haworth co-indexing entry note]: "The Implementation of Information Technology in the Corporate Engineering Library." Schwarzwalder, Robert. Co-published simultaneously in *Science & Technology Libraries* (The Haworth Information Press, an imprint of The Haworth Press, Inc.) Vol. 19, No. 3/4, 2001, pp. 189-205; and: *Engineering Libraries: Building Collections and Delivering Services* (ed: Thomas W. Conkling, and Linda R. Musser) The Haworth Information Press, an imprint of The Haworth Press, Inc., 2001, pp. 189-205. Single or multiple copies of this article are available for a fee from The Haworth Document Delivery Service [1-800-342-9678, 9:00 a.m. - 5:00 p.m. (EST). E-mail address: getinfo@haworthpressinc.com].

technological opportunity has resulted in a dramatic redefinition of the corporate library and its emergence as a novel information organization. Given recent technical advances and the expanding pace of corporate consolidation, it is expected that the corporate engineering library will develop a stronger focus on information technology in the coming decade. *[Article copies available for a fee from The Haworth Document Delivery Service: 1-800-342-9678. E-mail address: <getinfo@haworthpressinc.com> Website: <http://www.HaworthPress.com> © 2001 by The Haworth Press, Inc. All rights reserved.]*

KEYWORDS. Corporate engineering libraries, management, information technology applications

INTRODUCTION

Of what has been written on engineering libraries, the vast majority addresses the academic setting. While the literature is replete with papers discussing specific innovations at individual corporate libraries or in discussing challenges to corporate information centers, few provide an overview of the nature or focus of the corporate technical information center. A few papers do stand out as noteworthy in their broader coverage or exploration of mission. Hall and Jones (2000) examine Intranet innovations by corporate libraries through a series of case studies. Smith's (1997) survey of 118 companies provides insights into the effects of business trends on corporate libraries. Degoul et al. (1990) explore how corporate libraries could better integrate scientific and technical information with the business activities of their companies.

The lack of coverage of corporate engineering librarianship in library journals is not surprising since the majority of technical libraries available to the public are affiliated with universities. However, the role of the technical library or information center in the corporate world is a significant one both in number of individuals employed and in the scope of their responsibilities. The goal of this paper is to describe the commercial and business trends that have shaped corporate engineering libraries in the recent past and to demonstrate how these organizations have taken advantage of information technology to survive, and in some cases to thrive, in a difficult environment. An examination of how information technology has been utilized in corporate engineering libraries and the manner in which it has shaped their direction provides insights on organizational change and opportunities for libraries in other settings.

Corporate libraries are common in industries focused on pharmaceutical, chemical, and engineering disciplines. Unlike the majority of corporate libraries, which focus on business or financial disciplines and have one or two staff members, corporate engineering libraries tend to be larger and more elaborate in structure. Prior to discussing modern corporate engineering libraries, it is important to describe some of the factors that have shaped their development and to differentiate between corporate and academic engineering libraries.

In the corporate world there is a clear focus on profit and productivity. As a cost center in a profit-making enterprise, the library must be able to demonstrate a clear benefit to the organization in order to survive. Furthermore, the library must demonstrate the cost-effectiveness of its services and collections with respect to external options. This emphasis on efficiency and effectiveness has led many corporate libraries to embrace outsourcing as a means of addressing many processing and technical services functions (Brown-Woodson, 1998; Stear et al., 1996). By outsourcing operations–such as document delivery, cataloguing, interlibrary loan, and some database searching–that can be performed at acceptable quality levels at less cost, the library can reduce cost and focus staffing on activities that have greater business value (Olson, 1995). The development of the hybrid (insourced-outsourced) corporate library was a difficult task when initiated in the early 1990s (Stear and Wecksell 1996a, 1996b). However, it proved to be a useful pre-adaptation to the advent of Web-based information services and products and allowed the corporate library to develop a faster and more streamlined service response system than its academic counterpart.

An important aspect of the corporate library is the relationship between the collection and the library. In academia, libraries are primarily defined by their collections. Rankings of large academic libraries typically refer to the size of collections, as do university promotional materials when appealing to prospective students. Collection size and completeness have been objectives of the academic library since the Farmington Plan. Academic science and engineering libraries, which have a tendency to exist as branch libraries, have collections that are defined by the disciplines of their related academic units. This results in some degree of segmentation of collections by subject among libraries serving scientific, engineering, and business colleges or departments. Corporate engineering libraries tend to have small core collections that cover a wide range of subject areas. Focused on answering queries and serving no real historical or archival role, they place a strong emphasis on information access rather than information ownership. While academic libraries are typically evaluated by collection size, the performance of corporate libraries is measured in terms of services and the degree of integration with their customers.[1] Examination of budgets in the academic and corporate settings reflects this difference in mission. Corporate engineering library budgetary allocations to collections are far smaller on average than those of ARL Libraries, while they exceed academic library budgetary allocations in the areas of services, online access, and document delivery.[2] This difference in operational focus is reflected in the manner in which corporate and academic libraries have pursued the development of the digital (or virtual, or electronic) library and marks the divergence of corporate and academic engineering librarianship.

THE DEVELOPMENT OF THE DIGITAL CORPORATE LIBRARY

The advent of the Intranet was a watershed event for corporate libraries. The group of information technologies associated with the Web (digital content, hyperlinkage, access and convenient interfaces to remote databases, etc.) allowed the corporate engineering library to serve a widely dispersed audience and provide just-in-time information services. Those libraries that embraced new technologies and co-opted the Web as a part of their portfolio tended to prosper. Many of those who ignored the growth of the Web and digital information stagnated or were eliminated (Schwarzwalder, 2000a; Stear et al., 1996). Most corporate engineering libraries recognized the value of information technology and aggressively sought ways to implement it in their environments. As with their academic brethren, corporate Web developers' first efforts combined links to Internet information resources with information about their centers and services. Both academic and corporate library Web sites began to incorporate wider access to electronic library catalogs and internal and external bibliographic databases during the mid to late 1990s. The real divergence came in the late 1990s as academic consortia began to acquire large collections of electronic journals and corporate libraries built upon their access to internal corporate information and their experience with outsourcing arrangements to build service infrastructures.

Perhaps most influential in the evolution of digital corporate technical libraries has been their role in disseminating internal corporate information. Large corporations spend millions or billions of dollars per year on research and development. This work tends to be highly focused on the mainstream business interests of those corporations. While it is commonly assumed that this information is readily accessible within companies, historically this is far from true. Even in the early 1990s it was far easier to obtain external information than internal corporate information. That lack was the result of a variety of factors. Almost all companies had very inefficient mechanisms for the internal publication of information (Cote, 1995). The management of intellectual content was typically left to individual business units. While there were almost always corporate policies governing the management and retention of internal information, ensuring compliance was difficult.

Beginning in the early 1990s, corporations became more aware of the importance of internal information. Increasing product litigation required companies to submit as evidence experimental and design records that were often irretrievable. Courts tended to have very little sympathy with corporations that had poor record management programs and often punished them with steep fines and other punitive measures. Knowledge management gurus began to alert corporate administrators to the financial value of internal information. These trends triggered renewed interest in corporate information centers as possible solutions to the problems inherent in the management of internal information. In addition, the implementation of the ISO 9000 quality manufacturing standard created new requirements for documenting processes and providing access to that documentation. Corporate libraries seized upon digital publication methodologies and began managing repositories of PDF and HTML documents. Familiar with issues of information security, database and catalog management, indexing (metadata) and retrieval, they were well positioned to provide effective solutions to new corporate information mandates.

In addition to the need for better management of internal documentation, outsourcing was a major stimulus to the development of the digital corporate library. The role of outsourcing is a major issue for corporate libraries and sets the tone for a great deal of recent corporate library development (Stear et al., 1996). Outsourcing began as a threat to the corporate library in the late 1980s and early 1990s as corporations downsized and reduced staffs and budgets. Commercial library-services organizations replaced some libraries, but soon found that their most receptive audiences were the libraries themselves. Corporate librarians quickly responded to the threat of outsourcing by selectively outsourcing their own organizations (Brown-Woodson, 1998). Outsourcing operations such as document delivery, acquisition processing, and cataloguing had little effect on customer satisfaction and generated significant cost savings. Another wave of outsourcing hit corporate libraries in the mid 1990s with the growth of the Internet (Marcum, 1998). Initiatives by outsourcers and database consolidators to replace corporate libraries with Internet solutions hastened the demise of a small number of corporate libraries (Field, 1995; Kirchner, 1995; Stear and Wecksell, 1996b). However, most large corporate libraries had developed a strategy for embracing both the Web and selective outsourcing by that time.

The challenge of outsourcing has always been to integrate remote services with local services in a manner in which the distinction was invisible to the customer. While the drive to selectively outsource was a successful business strategy for the corporate engineering library, it was an administrative nightmare—until the advent of the Web. As information vendors adopted the Web, corporate libraries were able to develop interfaces between internal Websites and these external content and service providers. These early implementations of transactional processes provided corporate libraries with expertise that helped expand their roles in corporate information and knowledge management efforts. Experience with information interface design readily translated into the ability to design middleware to connect legacy information systems with the growing resources of the corporate Intranet. The advantages of this kind of development can be seen in development of a standards information system by Ford Motor Company's RLIS Group (Research Libraries and Information Services). The standards information system incorporates access to numerous collections of internal standards and specifications, searching and delivery of external information standards, access to full image versions of key industry standards and standards collections, and a function by which engineers who order a standard are notified when that standard is updated. The services are integrated with user data so that requests are routed to help centers located in the same regions of the world as the users. While the service relies upon transactions involving several external companies and several internal departments, the complexity of the operation is largely invisible to the user. The scope and complexity of the Ford standards information system would be extremely difficult to accommodate without external information providers who were proficient in information technology.

Both of these influences were of secondary interest in the academic setting. While academic libraries typically have experience with outsourcing some cataloguing functions to OCLC, outsourcing has not been a major trend in the academic setting (Bannister et al., 1999; Libby and Caudle, 1997; Osif and Harwood, 2000). Less experienced with extensive outsourcing, academic libraries have not been driven to develop the degree of process integration seen in the corporate world. Since publication in the academic setting needs to occur through academic journals in order to validate the credentials of the faculty, there has been little role for the library as publisher. Much of the energy in building "digital libraries" on the academic side has involved creating large collections of electronic periodicals. In most cases, this has involved the licensing of journal content or access by libraries or library consortia. In some cases, universities have been actively engaged in the digitization of older periodical literature. Corporate libraries were able to be fast adopters of digital journals, but have emphasized ownership of small (100-500) core journal collections with strong document delivery services. In the last few years, corporate libraries have stressed the development of services that deliver copies of articles electronically rather than on increasing the number of electronic journal subscriptions.

It is this integration of the corporate engineering library with the information producers and systems administrators in the corporation that marks the departure of the corporate information center from the role of library. The evolution of role from information provider to information intermediary has been the salvation of the corporate technical library in the cost-conscious

1990s and has dramatically expanded the value and opportunities of these organizations.

THE ROLES OF THE CORPORATE TECHNICAL INFORMATION CENTER

Information Provider

As with most library organizations, the traditional and primary roles of the corporate engineering library are in providing information to their clientele. While the academic library accomplishes this largely on the basis of a large physical or digital collection, the corporate counterpart relies more on rapid, commercial, document delivery services. Unlike the academic engineering library, the corporate center typically has a much more geographically dispersed clientele that relies more upon electronic communications with the library. There is a far greater range of services offered to corporate clientele, reflected in budgets that heavily favor services over collections, the inverse of the situation in academic settings. The corporate client base typically places a heavy premium on obtaining results that have been analyzed, summarized, and reported in brief. Required response time is typically short (days) to very short (hours) and there may be major financial consequences to incorrect or incomplete information. Internal information typically carries some degree of confidentiality and there is often segmentation within the ranks of the company regarding who is permitted access to which information. Internal information typically fulfills between 40% and 70% of the information requests to corporate information centers,[3] reflecting the extensive research and development activities of corporations in the engineering, scientific, and pharmaceutical industries. Besides meeting engineering demands, the library frequently also serves business, administrative, and legal interests.

The geographic distribution of the clientele and extensive use of remote services have predisposed the corporate engineering library toward the development of digital means of information dissemination. In order to address corporate needs, information systems required: the seamless integration of external and corporate information services; the integration of a hierarchical security system; the automated submission, routing, and receipt of service requests; the incorporation of legacy database systems; and, the provision of access to large collections of digital information. Corporate technical libraries' use of internal networks (Intranets) as a primary means of reaching their customers provided an excellent vehicle for the deployment of systems that met these objectives. The strength of the Intranet as a mode of asynchronous service delivery, as a flexible interface for the fusion of legacy systems and collections, and its provision of a security structure, allowed corporate libraries to turn their administrative burdens into service assets.

While most corporate engineering libraries have continued to operate reference and online search services, the number of people they serve by providing information and automated service options on the Intranet exceed those served through traditional services by orders of magnitude. The success of the library as digital service provider has led these organizations to explore additional means through which they could exploit this new medium.

Beyond expediting the manner in which corporate libraries have reached their clients, information technology has revolutionized the scope of services provided by these organizations. Through the use of current awareness services, such as Ford Motor Company's RLIS Select product (Figure 1), data mining, and text visualization technologies, libraries have been able to expand their horizons into public affairs and competitive intelligence functions. Involving information professionals in these areas results in the availability of a highly relevant stream of information to decision makers and allows them to focus their activities on core business needs–as opposed to information gathering. Moreover, advanced information technologies, such as data mining and visualization, move beyond retrieval and allow the information professional to identify trends and patterns within large sets of complex textual data. This often provides an entirely new capacity to corporations that can be applied to a wide variety of information streams. The role of corporate libraries in performing these types of analyses is emerging as an area of potential value.

Internal Publisher

A traditional, and often lamented, role of the corporate engineering library is that of archivist or records retention center. While many major corporations have separate archives, these departments most frequently concern themselves with corporate history, and the management of intellectual content is frequently left to the library. As better scientific and engineering text-editing packages became available in the 1990s, engineers and scientists abandoned the cut-and-paste method of technical report preparation and began producing totally digital technical communications. With the availability of digital technical reports, corporate libraries began efforts to make these reports available through the Intranet. The ability to merge internal technical literature of high relevance to their companies with external technical literature established the role of the corporate library as a content-rich information source and as a partner in the internal promotion of the R&D organization within the corporation. At Ford Motor Company we have expanded our role as a promoter of corporate R&D by providing a series of ancillary services that focus upon the promotion of internal R&D. Primary among these is our "RLIS Select" product,

FIGURE 1. Ford Motor Company's RLIS Select is a Web-based subscription service. Users select areas of interest. As new information is added to the system, articles of interest are automatically sent via e-mail to corporate subscribers. The system also allows users to submit their own information for dissemination to the RLIS Select user community.

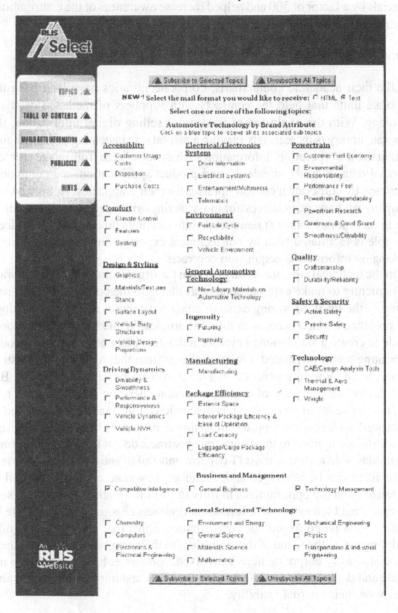

which sends e-mail alerts of new information resources in narrowly defined subject categories to profiled subscribers (see Figure 1). Through the integration of digitally published internal technical reports with the library online catalog and the RLIS Select service, we have increased the usage of these materials by a factor of 500 and helped increase awareness of the contributions of our research organization to the company as a whole.

Content Intermediary

Like their academic counterparts, corporate libraries have long been the corporate units that held the contracts with suppliers of mediated database searching. With the explosion of companies selling digital information, the corporate library is increasingly playing a pivotal role in helping to bring digital content to the corporation. Information such as news feeds, full-image articles, financial data, industry standards, and product specifications are commonly purchased by corporate technical libraries for their companies. At Ford Motor Company we have been successful in partnering with our Public Affairs division to incorporate external news feeds into internal corporate news services. This role is facilitated both by our technical expertise and our experience in working on information acquisition contracts.

On the technical side, the library is often the organization with the software infrastructure to make external content available to the corporation. This role is largely the result of having existing citation databases and library catalogs that provide enhanced access to full-text articles. While a corporation could decide to create a stand-alone keyword index to a body of literature, it would be creating a non-integrated, low-precision system that replicates an existing service. Such things are generally not done in cost-conscious companies. Beyond the technical aspects of content provision, there are strong business reasons to involve the library in contract discussions. Information as a commodity is not well understood by most purchasing departments. The standard terms and conditions applied to most corporate contracts do not translate to information products. In addition, most IT departments fail to anticipate how information content can be leveraged and, therefore, negotiate contracts that fail to capitalize upon the opportunities to fully utilize information products. In several cases that I am aware of, information contracts negotiated by corporate IT organizations failed to provide redistribution rights that would have significantly improved the value of those services to their corporations. Where there has been a savvy corporate library involved, companies have been able to negotiate and deploy superior information solutions and libraries have been able to increase their internal visibility.

Information Research Center

The corporate libraries at Hewlett-Packard and Ford Motor Company have extended the range of their activities by adding a research function. In Hewlett-Packard's case the emphasis is on the nature of the research process (Prime, 2000); at Ford (Schwarzwalder, 2000b), the emphasis is on information technology. In both of these examples, corporate engineering libraries have capitalized upon their roles to organically extend the ranges of their activities. For Hewlett-Packard, the integration of information professionals into the engineering process led to their interest in and work on the knowledge creation process. At Ford Motor Company, by identifying needs beyond the range of existing information technology, the library was able to add staff to pursue the development of new technical approaches to information analysis and dissemination. The addition of research responsibilities to a corporate engineering library already heavily involved in designing information solutions is a logical development. By integrating the functions of information provision, information systems design, and information technology research corporations can ensure research efforts that are targeted on pragmatic goals. Moreover, by linking the relevant research and development functions, companies can improve their ability to move applications from theory into practice.

FOSTERING TECHNOLOGICAL DEVELOPMENT IN THE CORPORATE ENGINEERING LIBRARY

Innovations in information technologies and the operational environment of the 1990s reshaped corporate technical libraries from institutions that resembled small academic libraries into hybrid library/information technology centers. The rapid development of corporate Intranets in the 1990s and the current enthusiasm over Business-to-Business (B2B) and Business-to-Consumer (B2C) applications has further rewarded those corporate libraries that took aggressive roles in the early development of Intranet and Extranet information resources. Traditional skills in information mediation have translated into expertise in interface design. Cataloguing skills have translated into record and system design insights. Years of interpersonal networking within the corporation to locate collections of obscure internal information have forged diverse partnerships and allowed corporate librarians to create the computer networks that link legacy information systems.

While many corporate library organizations have been remarkably successful in making this transition, it has required a variety of resources not traditionally associated with libraries and has involved a high degree of risk. This risk is due to the fact that meaningful information technology innovation requires tailoring to a corporation. Shrink-wrapped information products seldom, if ever, perform as promised. Libraries that have attempted to implement quick technical fixes have fared as poorly as central IT organizations that have taken the same approach. Successful innovation has required tailored solutions and real involvement with the technologies. Moreover, the transition from a traditional library to one that takes full advantage of information technology requires a change in the culture and operational design of the organization. The integration of an information technology focus into the mission of a corporate technical library is not a trivial matter. It requires administrative support, the creation of an organizational structure that supports innovation, and good strategic and tactical planning.

Administrative Support

The primary factor in gaining corporate support is to align funding proposals with the objectives of the corporation. While this seems like a trivial observation, most funding requests are steeped in language about why the author thinks the work is important. Proposals that support the objectives of local and central corporate executives are far more likely to be supported than proposals that have no basis in company priorities. By strategically positioning synergistic projects that meet specific corporate objectives to build upon selected technical approaches, a long-term development effort can be supported while achieving short-term successes along the way. The greatest danger in this approach is mission drift, the involvement in peripheral efforts that divert attention and resources from the primary mission of the organization.

Operational Structure of the Corporate Engineering Library

Traditionally, corporate libraries are designed around either functions or customer groups. A functional organization is typically hierarchically divided with staff groups responsible for activities such as online searching or information services. Customer-centric organizations are a more recent innovation and bring cross-functional staff together into groups that serve specific business units. Customer-centric approaches are common in corporate research environments. Both of these approaches are well suited to performing relatively stable work assignments, but are less appropriate for technological innovation. In contrast to either of these approaches, the assignment of cross functional, cross organizational project teams with an understanding of consumer needs and the relevant technologies is often the best approach for generating meaningful advances in information technology projects. A project team organizational structure allows the rapid deployment of staff to new projects, safeguards against parochial or sub-optimal solutions, and improves staff morale by providing people with dynamic and challenging positions.

While this project team approach is well suited to the development of new ideas, a new idea is not necessarily a successful idea. To gain the financial and staffing support necessary to launch a new information product, it is necessary to demonstrate its effectiveness, its scalability, and find a source of funding. Our approach at Ford Motor Company has been to develop a series of demonstration pilot projects, run them long enough to gain an understanding of the scalability and performance factors, and promote them to executives who have an interest in those functionalities. In any effort that is experimental in nature, there is a certain risk of failure. The topic of project failure is an important one that needs to be addressed openly. If objectives are ambitious, some efforts will not succeed. Evaluation of staff involved in failed projects should take into account the reasons for failure and the measures the staff members took to overcome difficulties. It is quite possible that a failed project was the result of unavoidable circumstances despite outstanding efforts of all involved. By keeping pilot projects small, the expense of failed projects can be minimized. By basing staff evaluation on performance, staff members can be encouraged to take acceptable risks.

Issues of staffing are complex. Beyond developing a mix of people with librarian and programming skills, it is important to create synergies and foster collaborations. Besides their professional skill sets, people bring a variety of perspectives to the job. Some see issues in broad perspective; others have a superb grasp of details. Some are ruthlessly analytical; others, hopelessly artistic. A mix of skills, perspectives, abilities and focus can create something impossible to achieve from a single mind. The creation of an environment that values intellectual diversity is essential to innovation and creativity, but is far easier to discuss than to implement. In all work settings, there is a price to be paid for intellectual openness, in terms of efficiency losses and personal ego, that few are willing to pay. Exploration of options and ideas takes time and effort and experimentation does not always lead to success. Preserving this type of environment requires a great deal of work, but is well worth the effort.

Strategic and Tactical Planning

In my experience, many information professionals seem confused about their real mission. They confuse the tools of their trade and the operations of their organization with their focus and mission. If the goal is to create a seamless integration of people and ideas, then staffing an information desk is, at best, an imperfect means of meeting that goal. If the goal is running a library, it may be a primary focus. The rapid pace of change in information technology has resulted in chaos in most library environments. New technologies are evaluated on the basis of how they support, or detract, from existing services. Electronic journals were seen as an asset by academic libraries because they could be collected in'the same manner as paper journals. The price differential between the electronic and the paper was a plus, because it helped differentiate between the top-notch libraries and the second tier players. Internet search engines have been seen as a threat by many libraries as they suggest to users that information access does not require the library as middleman. To engage in this type of short-term tactical thinking in the face of rapid technological change can leave libraries ill-prepared to face the future. A strategic view raises the question of what role the library should play. In Ford's case, that role involves integrating internal and external services, content, and technology, to link people and information. Seen in this light, external search engines become an asset to be co-opted and electronic journals become an ineffective artifact of a publication process. Most end-users want information that is embedded in journal articles. Buying a journal to get a section of an article is analogous to buying a car to get a cigarette lighter.

Understanding strategic focus is essential to developing short-term goals and selecting appropriate projects. Without a strategic focus, library managers will assemble an incoherent selection of projects that involve them in the latest information technologies, but do little to build strength or direction. This is a common problem across the profession at this time.

THE ROLE OF INFORMATION TECHNOLOGY IN THE CORPORATE ENGINEERING LIBRARY

The developments of the last decade have resulted in the wide-scale implementation of information technology in the corporate technical library. The resultant recognition of the value of the corporate library has created opportunities that continue to shape the directions of these organizations. While there will continue to be support for traditional library services within the corporate environment, there is a clear impetus for corporate engineering libraries to continue evolving a stronger technology focus. It is important to note that this is not an abandonment of core values. In fact, it is through the adoption of new information technologies that corporate engineering libraries are now able to fulfill the idealistic visions of the pioneers of modern librarianship. As corporations continue to strive towards globalization and maximum efficiency, corporate engineering libraries need to embrace technologies that allow twenty-four hour global access. There is no corporation that would support the proliferation of physical libraries to meet the information needs of a global organization. Beyond the efficiency gained by building a networked information resource, the corporate library must operate in this environment in order to be relevant to its customers.

The Intranet has become the primary means of communication for many large corporations. Experimentation with virtually collocated teams and telecommuting increasingly shifts the workplace from the office to the network. To be relevant in this new business environment, the library must be present in the virtual workplace. As new generations of computer-aided-engineering tools emerge, information must be integrated to work process in a manner that ensures accuracy and currency. As information volume increases, the library must move from information retrieval to information analysis. By helping filter, visualize, model, and analyze information, the information professional can help find patterns and trends within complex sets of information. In a number of applications, these trends may be inherently more valuable than the information itself (e.g., business trends for competitive intelligence customers). By linking information services to business needs, the corporate technical libraries can drive changes that provide real competitive advantage for their corporations.

In the last ten years the corporate technical library has progressed from technical obsolescence to the status of early technology adopter (Stear 1997; Stear et al., 1996). In an increasing number of cases, the corporate library has become the organization that brings new information into the company. In a few cases, it is the library that helps create that technology. The corporate emphasis on financial success and the increasingly competitive global business environment suggests that this is not just the best course of action, but that it may be the only course of action.

Given the increasing role of information technology in the corporate technical library, the next decade will create some interesting examples of how these technologies can be applied in a library setting. Since we face a number of emerging information technologies, it is important that lessons learned in the corporate setting be communicated in the journal literature. Assessments of the values of these technologies, examples of how they can be applied, and the organizational implications of technology application in the corporate engineering library are all topics that have broader implications to the profession.

NOTES

1. The use of collection size to evaluate academic libraries is confirmed by the often-cited Chronicle of Higher Education survey on library holdings (2000), by the use of library holdings in popular university rankings (Peterson's, 1997) and through personal communications with Mark Sandler, Head of Collections for the University of Michigan Libraries. Information on evaluation metrics in corporate engineering libraries is derived from personal communications with other corporate library managers and from personal experience.

2. Retrospective ARL statistics (Kyrillidou and O'Connor, 2000) demonstrate that an increasing percentage of ARL library budgets are devoted to collections. In 1986 ARL libraries devoted an average of 32.3% of their budgets to collections and that figure has gradually increased to 37.4% in 1999. Total operating expenses, which include service expenses, have decreased from 13.5% to 12.8% over the same time period. An initial review of a recently completed budgetary survey of major corporate technical libraries reveals that $1.40 is spent on services for every $1.00 that is spent on collections (unpublished survey).

3. This figure is based upon personal experience and personal communications.

REFERENCES

Banister, Stephen, Sheley, Marie and Lee, Crystal. "Outsourcing in Louisiana's Academic Libraries." *LLA Bulletin*, vol. 61, no. 4 (1999): 212-216.

Brown-Woodson, Ina. "Online Services to AT&T Employees." *Library Trends*, vol. 47, no. 1 (1998): 172-179.

Chronicle of Higher Education. "Holdings of University Research Libraries in the U.S. and Canada, 1998-99." vol. 46, no. 37 (May 19, 2000): A23.

Cote, Stevie. *Ford Information Management: Records Management Reengineering Project Report* Part I: Current State Analysis. (April 27, 1995).

Degoul, P., Debrun, U., Ferrari, T., Werner, E. and Heydon, J. "Economic, Scientific and Technical Information (EST) for the Firm." *Infomediary*, vol. 4, no. 3/4 (1990): 139-157.

Field, Judy. "Downsizing, Reengineering, Outsourcing and Closing: Words to Lose Sleep Over." *Library Management Quarterly*, vol. 18, no. 4 (1995): 4.

Hall, Hazel and Jones, Alyn M. "Show off the Corporate Library." *International Journal of Information Management*, vol. 20, no. 2 (2000): 121-130.

Kirchner, Russell. "Save 20-50%–Outsource Your Corporate Library in 1995." Marketing letter sent by Russell Kirchner of Teltech Knowledge Management Services to executives at major U.S. corporations. (June 19, 1995).

Kyrillidou, Martha and O'Connor, Michael (Compilers and Editors). *ARL Statistics 1998-99*. Washington, D.C.: Association of Research Libraries, 2000.

Libby, Katherine, and Caudle, Dana. "A Survey on the Outsourcing of Cataloguing in Academic Libraries." *College and Research Libraries*, vol. 58, no. 6 (1997): 550-560.

Marcum, James W. "Outsourcing in Libraries: Tactic, Strategy, or 'Meta-Strategy'?" *Library Administration & Management*, vol. 12, no. 1 (1998): 15-25.

Olson, Christopher. "Outsourcing: Opportunity or Threat?" *Marketing Treasures* (September/October 1995): 4-6.

Osif, Bonnie and Harwood, Richard. "Privatization and Outsourcing." *Library Administration & Management*, vol. 14, no. 2 (2000): 102-107.

Peterson's Guides, Inc. *Peterson's Guide to Four-Year Colleges*. 27th ed. Princeton, NJ: Peterson's Guides, 1997.

Prime, Eugenie. "Research in a Research Library." *Dynamic Systems and Services: Special Needs, Special Libraries*; A Panel in a Series hosted by the University of

Michigan School of Information in their "Library Cultures" series. (*http://www.si. umich.edu/library-cultures/special/corporate/prime.ppt*) (March 29, 2000).

Schwarzwalder, Robert. "Manifesto: Seizing the Initiative in the Information Economy." *Econtent*, vol. 23, no. 1 (2000a): 60-63.

Schwarzwalder, Robert. "Reinventing the Library at Ford." *Dynamic Systems and Services: Special Needs, Special Libraries*; A Panel in a Series hosted by the University of Michigan School of Information in their "Library Cultures" series. (*http://www. si.umich.edu/library-cultures/special/corporate/schwarzwalder.ppt*) (March 29, 2000b).

Smith, Norma K. "The Effects of Business Trends on Corporate Libraries: Science and Engineering Libraries with Holdings in Polymer Science." *Science & Technology Libraries*, vol. 17, no. 1 (1997): 53-66.

Stear, E. "Aligning the Corporate Library With the Business at Digital." *Gartner Research Note.* (November 13, 1997).

Stear, E. and Wecksell, J. "IRC Outsourcing: Competitive Threat or Technology Trend?" *Gartner Research Note.* (September 27, 1996a).

Stear, E. and Wecksell, J. "IRC Survey Results Identify Issues of Most Importance." *Gartner Research Note.* (October 1, 1996b).

Stear, E., Wecksell, J. and Bosik, D. "The Evolving Role of Information Resource Centers." *Gartner Strategic Analysis Report.* (September 26, 1996).

Focusing on the User
for Improved Service Quality

Deborah Helman

Lisa R. Horowitz

SUMMARY. With the recent explosion of the World Wide Web, people now have many options when seeking information. They are much less likely to look first to the wealth of resources and services that libraries offer. One of the greatest challenges for libraries today is to re-establish ourselves as the first and foremost gateway to which our users turn for assistance in navigating through the vast amount of information at their fingertips. A key to our success will be a focus on our users, proactively assessing their needs and responding with quality service that meets those needs–before they find the need to go elsewhere. This article will describe what MIT Libraries Public Services have done to address this challenge over the past three years as we directly confronted issues of service quality and performance measurement. Two key outcomes that will be discussed are the development of user groups to focus on users and their issues, and a service philosophy statement that is the basis for the creation of a strategic plan for quality service. *[Article copies available for a fee from The Haworth Document Delivery Service: 1-800-342-9678. E-mail address: <getinfo@haworthpressinc.com> Website: <http://www. HaworthPress.com> © 2001 by The Haworth Press, Inc. All rights reserved.]*

Deborah Helman, BS, MLS, is Associate Head and Reference Coordinator, Barker Engineering Library, Massachusetts Institute of Technology. Lisa R. Horowitz, AB, MA, MLS, is Foreign Languages Librarian and Data Specialist, Humanities Library, Massachusetts Institute of Technology.

The authors gratefully acknowledge the help and input of the individuals who contributed content or reviewed the paper for accuracy and clarity: Ginny Steel, Eileen Dorschner, Michael Finigan, Michael Leininger, Nora Murphy, and Sarah Wenzel.

[Haworth co-indexing entry note]: "Focusing on the User for Improved Service Quality." Helman, Deborah. and Lisa R. Horowitz. Co-published simultaneously in *Science & Technology Libraries* (The Haworth Information Press. an imprint of The Haworth Press, Inc.) Vol. 19, No. 3/4, 2001, pp. 207-219; and: *Engineering Libraries: Building Collections and Delivering Services* (ed: Thomas W. Conkling, and Linda R. Musser) The Haworth Information Press, an imprint of The Haworth Press, Inc., 2001, pp. 207-219. Single or multiple copies of this article are available for a fee from The Haworth Document Delivery Service [1-800-342-9678, 9:00 a.m. - 5:00 p.m. (EST). E-mail address: getinfo@haworthpressinc.com].

KEYWORDS. Customer service, customer service philosophy, performance measurement, service quality

INTRODUCTION

Thirty years ago people seeking information came to the library because libraries were repositories for knowledge that had been compiled in print. The World Wide Web did not exist. There were few if any alternatives for library users to find information, so libraries did not have to be strategic about providing stellar service and instead focused on building large collections. Now, however, for students entering college at the beginning of this century, the Web is second nature, and it is growing quickly into a primary resource. Today people have many options when looking for information and often do not choose the library as their first place to find information. How do we re-establish libraries as the first and foremost gateway to which our users turn for information?

Libraries need to address this major challenge by proactively assessing our services and their role in answering our users' needs and expectations, and by developing strategic plans for service quality. We must focus on our users, anticipate their needs and respond–*before* they find the need to go elsewhere. At the same time, we must assess our traditional programs and services so that we can consciously discontinue low impact services. It is imperative that we allocate our resources efficiently and justify new resources at a time when resources are limited. If we are successful in strategic planning to enhance service quality, we anticipate that this will encourage people to use the library when they are searching for information because they will be confident that they will receive consistently reliable service that may exceed their expectations.

This article will describe what the MIT Libraries Public Services have done to address this challenge over the past three years as we have directly confronted issues of service quality and performance measurement. We began with a process of redefinition that would articulate a plan for service quality in a user-centered organization. As a key part of that process a task force outlined measures for assessing service quality. Two key outcomes of this process which will contribute to assessment: user groups to focus on users and their issues, and a service philosophy statement that will become the basis of a strategic plan for service.

THE MIT LIBRARIES' EXAMPLE

Redefinition

In 1998, MIT Libraries Public Services began a process to plan for world-class service for the future. The Associate Director for Public Services, who had been at MIT Libraries for less than six months, reviewed feedback she had been gathering from Public Services staff while familiarizing herself with the Libraries, and realized that a major organizational "redefinition" was critical to the future of the MIT Libraries. She noted a number of trends in the comments she was receiving from staff:

- With the expansion of the "digital" library, all staff had been expected to take on more responsibilities, while still maintaining our traditional services. Job descriptions, however, had not changed.
- We were expected to provide new and traditional services with little or no increase in resources.
- In this changing environment, "our organizational structure [did not] allow us the flexibility and autonomy to respond quickly"[1] to meet our users' needs.
- Due to the organizational hierarchy, staff did not feel empowered to take action that might fill a user's need.
- Training was inadequate for the new skills staff needed to perform well in this digital environment.

As a result of this feedback, the Associate Director for Public Services initiated a process of redefinition in which all library staff were invited to take part (referred to as "Redefinition"). Task forces were formed to explore issues, with the ultimate goal of establishing Public Services that are enabled to provide world-class service for the future. The task forces worked in sequence to examine our organizational values, service priorities, organizational structure, internal communications and performance measurement, while a final task force provided constant communication about Redefinition. Each task force focused on ways to create an organization that could excel in supporting library users of the twenty-first century.

Assessing library services must begin with a review of those issues. Staff must come to an agreement on the library's values to form a foundation upon which services and assessment are based. Establishing priorities then allows the organization as a whole to focus on what is valued most by users. In turn, the structure of the organization must support the priorities and the goal of having a user-centered organization. Excellent communication–which includes decision-making, general knowledge about library activities, and committee structure– then allows information to be shared throughout that organization. Assessment benefits from good organizational communication. Focusing on performance measurement itself means that values and priorities can then be incorporated in the plan for assessment. Public Services went through this process, identifying values, service priorities and a logical organizational structure, and then began to study in depth how to assess new and established programs.

Performance Measurement Task Force

As part of the Redefinition, one task force was charged, "To review our current practices for measuring organizational efficiency and effectiveness, and to recommend a methodology to assess our performance based on the results of the Redefinition process." The Performance Measurement Task Force's (PMTF) first step was to define performance measurement, and put it into the MIT context. After conducting a thorough literature search, the definition that the PMTF believes met the needs of the MIT Libraries the best was Roswitha Poll's:

> Collection of statistical and other data describing the performance of the library, and the analysis of these data in order to evaluate performance. Or, in other words: Comparing what a library is doing (performance) with what it is meant to do (mission) and wants to achieve (goals).[2]

The task force met for 10 weeks, evaluating and analyzing performance measurement in libraries by reviewing the literature and having discussions with people at other institutions. Working with an MIT performance consultant, they explored performance measurement beyond the library context as well. Separately, staff representing all departments in Public and Collections Services were interviewed to compile a chart of the kinds of raw data collected in the MIT Libraries that might have an impact on Public Services. When discussing and integrating the gathered information, the task force determined that a single, clearly articulated customer service program (now simply a "service program") was needed to create a "culture of assessment"[3] at the MIT Libraries, which would ultimately be the basis for a new organizational culture. A well-defined service program would provide a context for performance measurement and would cultivate buy-in from Public Services staff, who would be the creators of the program. Everyone on staff would be involved in identifying and compiling data to be gathered that would contribute to the assessment and thus to the service quality of the Libraries.

A user-focused service program would include the following elements:

- "a written statement of customer service philosophy;
- training for [staff] in effective service delivery . . . ;
- measurement of service quality and consequent adjustment of service policy and delivery;

- use of data to adjust policies, services, or operations to better serve customers;
- organized processes for gathering data about customer behavior and satisfaction;
- service policies that provide latitude for staff as they serve customers . . ."[4]

The service program would include identifying needs and expectations of specific and potential user groups to respond to changing user needs. It would also provide a structure for obtaining a range of feedback from current users–remote and on-site–and potential users via web-based comment forms, suggestion boxes, focus groups, and other methods, enabling the Libraries to verify findings and determine service needs.

The PMTF identified five requirements for a valid service program, which we now consider "success criteria," that set of criteria against which we measure ourselves. The five criteria are that the service program:

- Values user expectations
- Includes all stakeholders
- Is proactive, ongoing, and flexible
- Ensures confidentiality
- Is based on sound theory

Performance measurement and the PMTF recommendations were new concepts to MIT Libraries staff. We had been collecting quantitative data–reference statistics, door counts, circulation statistics, etc.–and using it to some extent. We had yet to fully absorb what it meant to have a "culture of assessment." At the same time, as the Libraries began implementing the recommendations from Redefinition, we realized that the year of Redefinition had already begun to evolve our organizational culture. People were unconsciously incorporating some of the philosophies that came out of Redefinition, such as the continual need for user feedback and more meaningful statistics, into their daily activities.

Two things were still missing though, as noted in the PMTF recommendations: (1) qualitative data with a structured mechanism for gathering ongoing user feedback, and (2) an overarching plan for service quality. These gaps drove the first steps of the implementation process for the recommendations that came out of Redefinition. The Libraries established user-focused committees, which we refer to as user groups, to address the need for ongoing user feedback. We also began development of the service program that would eventually serve as the foundation for a comprehensive plan for service quality.

USER GROUPS

User groups in the MIT Libraries were established through the Redefinition process and would ultimately become part of the service program. The idea behind the user groups was to expand the tradition of having committees and administrative groups formed for handling and managing issues related to job functions, and instead to broaden the constellation of groups to focus on users and their issues. Previously, the Libraries focused on users solely by subject/department through subject specialists–bibliographers/reference librarians who act as liaisons to each department–but there had never been a library-wide means for assessing needs of the different groups that make up a department's constituency based on user category. There had been no systematic and ongoing way to evaluate library services, whether we provided them or not, that a particular user community might value, other than by subject. Also, the Libraries did not always have relationships with MIT's administrative groups that represented the various user communities.

The user groups changed these patterns. User groups are a systematic and ongoing means for the Libraries to connect to our user communities and the official MIT administrative groups that represent them, while bringing back knowledge of those communities' library needs. The user group members thus become a part of their community, and they act as a vehicle for communication for that community to the Libraries. They also act as champions on behalf of that user community, making recommendations to improve services.

The structure of MIT led to the formation of seven user groups: faculty, graduate students, undergraduate students, researchers, alumni, administrative staff and outside users. Each group, comprised of librarians and professional library staff, is led by a different department head in Public Services. Every librarian and professional staff member in Public Services chose a user group, with attempts to diversify the groups across libraries and subjects so that each group could focus broadly on the specified users.

At the time of this writing, five of the user groups have been active for almost a year, and a sixth is just starting. Each user group is analyzing a variety of issues to gain an understanding of their own users' needs. To begin, each group wrote a mission statement that clarified the purpose of the group for its members and other library staff. Some of the groups also identified specific goals and activities for assessing users' needs, as well as possible enhancements to currently offered services.

The first step of each user group was to define the user community that they serve. For example, the Administrative User Group understands that not only the governing administration (such as the President's and Chancellor's offices) would be included in this group, but also other staff, such as departmental staff assistants, human resources staff, and staff in the museum and the dining facilities. Because the user groups are vehicles for assessment and communication, each group must also identify the mechanisms by which to disseminate its findings and delegate responsibility for action. For instance, the Faculty User Group realized that in order to have any impact, the survey they conducted of faculty would have to be reviewed by the functional committees (circulation, reference, collections, etc.), who would need to take an active role in creating new and improved services based on the results of the survey.

Each group has also begun to determine what services were already available to these groups, and to package that information in a systematic way. Some of this information had never been described before, although the services existed. For the Alumni User Group, creating a web page with resources focused specifically for MIT alumni was a relatively simple and helpful way to begin this process. They explored college and university web sites across the country to determine what is offered to alumni elsewhere. They discovered that there were few if any web sites describing services to alumni, so they invented their own structure. The new pages for MIT alumni have proven not only useful to alumni, but also to staff for answering questions asked by alumni. Another example of this is the work the Faculty User Group has undertaken to target new faculty. They are creating a welcome letter and packet that all subject specialists can use to orient new faculty in their departments to services of interest. The repercussion of that simple packaging of services in a clear way has been better service overall.

Assessing user needs is a high priority for each user group. Several groups have begun this process, in particular those groups whose constituencies are easier to define. The Graduate User Group has begun evaluation of data that was gathered by the Libraries in 1998. Their intent is to use this data to determine what the Libraries could change that would make it easier for graduate students to do their work. They are also discussing the survey results with the Graduate Student Council to make sure that the needs have not changed. Another example of assessment is the Faculty User Group survey, which was sent to all the faculty on campus. The resulting 27% response rate was considered successful. The user groups act on feedback, identifying action items and bringing them to the attention of the appropriate library staff system-wide.

Another element each user group emphasizes is outreach. User groups identified opportunities for interacting with their user community to enhance communication as well as formal and informal feedback. The Graduate User Group solidified their relationship with MIT's Graduate Student Council by agreeing to write a regular column about the Libraries in their monthly Graduate Student Council newsletter. Also as a result of this relationship, the Libraries hosted an open house for graduate students, held in the Barker Engineering Library, that attracted over 300 incoming graduate students. This informed incoming graduate students of our interest in their needs and our willingness to listen to their issues during their first week here. In another example, the Alumni User Group, with their new liaison to the MIT Alumni Office, has positioned the Libraries to gain feedback and to piggyback onto any surveying of alumni done through that office. And lastly, members of the Faculty User Group are regularly attending MIT-wide Faculty meetings.

The Administrative User Group will be attempting a slightly different mechanism for maintaining contact with their constituency. Each department on campus has a library liaison (their subject specialist) assigned already, so the group hopes to use these liaisons to inform administrative users about library services. The belief is that this would be particularly true for departmental staff assistants who often get sent to the Libraries but do not know how to find the information they need. As another way to keep users informed about services, the Administrative User Group is considering the possibility of including a library introduction in the orientation of new staff offered through the MIT Benefits Office.

This new structure of user groups has been one of the most exciting outcomes of the Redefinition because the efforts have involved almost all librarian and professional staff in the Libraries' Public Services, and, as a result, we foresee a major impact on service quality in the future. The concept of user groups was one of the first steps we took towards improving service quality and focusing on our users. It provided us with the missing mechanism for soliciting ongoing feedback from our users, creating an environment in which the MIT Libraries could become user-centered as an organization. Next we needed to develop the full service program mentioned above, which would become the foundation for an organizational culture of service quality in MIT Libraries Public Services.

SERVICE PROGRAM

As part of the implementation of the Redefinition, a newly formed committee, the Public Services Coordinating Committee (PSCC), composed of representatives of the Libraries' functional groups (reference, instruction, circulation, and processing) and Public Services administration, including both support staff and professional staff, took on the task of leading Public Services staff in the development of the service program. As they began to discuss how to develop the full service program, they identified a number of disparate elements that needed to be brought together: (1) the recommendations for the service program; (2) recommendations from Redefinition that were already implemented, such as the user groups; and (3) the new characteristics of our organizational culture assimilated during Redefinition. A member of MIT's Organizational & Development Team recommended that we use Leonard Berry's

"Framework for Great Service." Berry's "framework for great service" includes 6 major steps for creating great service in your organization: (1) nurture service leadership; (2) build a service quality information system; (3) create a service strategy; (4) implement the service strategy through structure; (5) implement the service strategy through technology; and (6) implement the service strategy through people.[5] PSCC members began by drafting a service philosophy for Public Services, as recommended by the PMTF, to provide a context within which to build the service program using Berry's framework.

The service philosophy would be the driving force of the service program, articulating a shared vision of world-class service in the libraries, for both staff and users. World-class service calls for two fundamentals: (1) an organization with a service-quality culture, and (2) an appreciation of users' expectations of service. Berry et al. explain, "Excellent service can . . . be energizing because it requires the building of an organizational culture in which people are challenged to perform to their potential and are recognized and rewarded when they do."[6] Hernon et al. describing service quality in a different context say, "Service quality does not deal with wants and needs, rather it focuses on expectations."[7] Creating a service quality culture will create an environment in which the MIT Libraries can meet and exceed the expectations of our users.

The process of developing the service philosophy began with customer service workshops with all Public Services staff. The objectives of the workshops were to have staff: (1) explore the concept of customer service; (2) discuss how the concept applied to libraries in general; (3) discuss how the concept applied to individual staff members' jobs; and (4) engage staff in identifying ways to improve service quality in their own jobs and empower them to act on these ideas. These workshops resulted in rich discussion and gave the PSCC a great amount of staff input to use in addition to the values identified during Redefinition.

An interesting result of the workshops was the ultimate decision of the PSCC to drop the use of "customer" that some staff had begun to use and still occasionally do. Many staff expressed great concern about the move toward the use of "customer" and "customer service." This concern has been noted in the literature as resistance to service quality. "Staff may not embrace the concept of customer service or service quality, because . . . , they claim that the concepts apply only to the private sector and businesses."[8] We also noticed that, as Hernon et al. predicted, "for some library staff, the very word 'customer' and a focus on 'customer service' are seen as a drift away from core values . . ."[9] In order to curb this reaction, the PSCC decided to not use "customer."

Based on the results of the workshops and following staff input from previous drafts, in Summer 2000 the PSCC finalized a service philosophy for Public Services:

We see ourselves as teachers and learners; we consistently challenge ourselves and welcome challenges from our users. We use our skills to anticipate services from which our users would benefit; together with our users we develop and refine our services considered most important. We develop and use standards to determine our goals and measure the quality of our services in pursuit of continuous improvement. We foster an environment where we treat everyone with respect and consideration. We empower staff to be innovative and to use their strengths and flexibility to provide our users with excellent service. We succeed when every user has successful encounters with MIT Libraries staff and resources.[10]

With the establishment of the service philosophy, MIT Libraries Public Services faces its greatest challenge–using it to inspire a service quality culture. As anyone who works in a large organization knows, a shift in culture is always difficult to effect. For Public Services staff to assimilate the service philosophy, staff will need to begin discussing the concepts involved. And, as the service program is developed in more depth with the service philosophy as its foundation, staff will have more opportunity to be exposed to the service philosophy and assimilate it into their daily work life.

FRAMEWORK FOR GREAT SERVICE

The development of the service philosophy laid the groundwork for the next steps of building the service program within Berry's "framework for great service." PSCC began by rating MIT Libraries Public Services against each one of these elements. They asked, "How are Public Services already doing with each one of these elements?" They were pleased to find that in terms of nurturing service leadership, building a quality information system and implementing service strategy through technology and people, Redefinition itself had initiated changes that meant Public Services were progressing in these areas. On the other hand, PSCC did find that Public Services were not being strategic about how we provided service. The service strategy, then, was a critical focus for us in order to improve service quality to our users.

Berry discusses four ways in which to nurture service leadership:[11] (1) "promote the right people"; (2) "stress personal involvement"; (3) "emphasize the trust factor"; and (4) "encourage leadership learning." PSCC found that we could apply these concepts to what had already happened as our organizational culture began to change. People from every library and every position in Public Services were invited to participate on each task force during Redefinition, and leaders for task forces were chosen from both the professional staff and the support staff. Support for that participation came from the highest levels in Public Services administration. At all stages of Redefinition, input and participation was sought from all staff. Staff on task forces were entrusted and empowered to make appropriate decisions based on the input gathered. As we implemented the recommendations, a systems model for decision-making[12] was developed to help move decision-making to lower levels, while reassuring management that stakeholders have appropriate input; this model is being reviewed for implementation within Public Services. During Redefinition, many leadership training opportunities were provided, with courses such as Creative Thinking Skills, Leading and Managing Change, Mastering Meetings, and Project Management. The initiatives to nurture leadership continue.

Performance measurement was instrumental in our progress in building a service quality information system. Berry states that "a service quality information system *uses multiple research approaches to systematically capture and disseminate service quality information to support decision making.*"[13] One recommendation of PMTF that has been implemented was to designate a member of each Public Services committee (e.g., circulation, instruction, processing, reference) as a champion of performance measurement, to guide and advise the committees in assessing services. The user groups described above also provide a conduit for "systematic listening"[14] to our users, revealing users' priorities and preferences for service, and guiding our decisions.

Within the context of the "framework for great service," there were several concepts that were undoubtedly beginning to permeate our organizational culture. However, we really had no all-encompassing strategy for service in the MIT Libraries Public Services. A service strategy is a useful tool because it provides "a meaningful, compelling, clearly articulated definition of what service means in your organization. It creates a vision of service excellence, clarifies service standards, and builds a service culture. Without it, people are left to their own interpretation of customer service which leads to inconsistency and dissimilarity."[15]

Berry defines the four elements of service strategy as "service reliability"[16] (that a user knows what to expect from a service), "service surprise"[17] (the "wow" that someone gets upon using a service), "service recovery"[18] (the response to a problem with service), and "service fairness"[19] (the overall sense that users have been treated fairly). Although we were certainly providing some of the elements of a service strategy—for example, we had fairly reliable service and our services were fair—we had no strategic plan for doing it consistently and across libraries, and we were not publicizing service standards to users, or even to staff. Our next goal in the continual evolution of Public Services is to articulate a service strategy for Public Services for our staff and our users.

CONCLUSION

Over the past three years MIT Libraries Public Services has worked to change our organizational culture to incorporate the ideals of service quality with a user focus. The articulation of our values, our service priorities, and our new communication and organizational structures (such as user groups) has been our foremost accomplishment thus far because it has unified the Public Services staff around a foundation upon which to build the service program. However, articulating these does not mean that people have fully incorporated them into their daily work life. In fact this has proven to be one of the greatest challenges to date.

At the time of this writing, that service program lives almost entirely in a single committee (PSCC) and still needs to be shared and then absorbed into the whole culture, but moving it outward is an awesome task that is still waiting to be tackled. A major obstacle has been how to reinforce these articulated ideals to current staff in a way that will result in meaningful changes in the way we work and, ultimately, in service quality to our users. Another issue that we face, as many other large organizations do, is the turnover of staff and how to communicate these ideals to new staff in a systematic way. Maintaining momentum for the larger process of improving service quality is also an enormous challenge in the face of all the other work and initiatives in which staff are involved. Finally, this initiative rests within Public Services, entirely separate from the other two departments in MIT Libraries, Collections Services and Systems, making change in organizational culture much more complex.

As we begin the next steps of this evolution, we must ensure that all staff members are given the opportunity to identify ways in which to live these new ideals and our new service philosophy. As we build our new service strategy, we believe that this understanding will lead to a staff that can fully contribute to the building process and support the service strategy as adapted over time. Yet as hard as it has been for us to incorporate what we have done so far into our daily work life it may prove even more difficult to incorporate the service strategy. As Barbara Wirtz has eloquently put it, "... the only thing harder than developing your service strategy is living it."[20] We want the MIT Libraries Public Services to live it and we must identify ways to ensure that this happens.

In building this service program, we hope to create a world-class service environment and a set of services that exceed our users' expectations and bring them back for more.

ENDNOTES

1. In an e-mail message that Ginny Steel sent to all Public Services staff on January 21, 1998.

2. Roswitha Poll, *Measuring Quality: International Guidelines for Performance Measurement in Academic Libraries*, IFLA publication, vol. 76 (Munich: K.G. Saur, 1996): 16.

3. At a visit to MIT, Amos Lakos of University of Waterloo introduced this term to us.

4. *Customer Service Programs in ARL Libraries*, SPEC Kit, vol. 231 (Washington, DC: ARL, 1998), 3.

5. Leonard L. Berry, *On Great Service: A Framework for Action*, (New York: The Free Press, 1995), 5.

6. Leonard L. Berry, A. Parasuraman, Valarie A. Zeithaml, and Dennis Adsit, "Improving Service Quality in America: Lessons Learned: An Executive Summary," *Academy of Management Executive* 8, no.2 (1994): 32-52.

7. Peter Hernon, Danuta A. Nitecki, and Ellen Altman, "Service Quality and Customer Satisfaction: an Assessment and Future Directions," *Journal of Academic Librarianship* 25, no.1 (1999): 9-17.

8. Ibid.

9. George Soete, *Customer Service Programs in ARL Libraries*, SPEC Flyer 231 (Washington, DC: ARL, 1998), 2.

10. Service Philosophy of MIT Libraries Public Services, Summer 2000.

11. Berry, *On Great Service: A Framework for Action*, 16-31.

12. Proposed "Systems Model for Decision-Making" from MIT Libraries Public Services, 1999. Retrieved 21 Oct. 2000. http://macfadden.mit.edu:9500/imps/handoff/decision-making.html.

13. Berry, his italics, p. 33.

14. Ibid.

15. Barbara Wirtz, "A Service Strategy: From Cliché to Action," *Marketing News* 34, no.10 (2000): 26.

16. Berry, p. 78.

17. Ibid., p. 89.

18. Ibid., p. 94.

19. Ibid., p. 108.

20. Wirtz, p. 26.

Opportunities for Creativity:
Fundraising for Engineering
and Science Libraries

Joanne V. Lerud
Lisa G. Dunn

SUMMARY. Fundraising for the academic engineering/science library must be an integral part of its operations in this time of restricted resources and rising expectations. Institutional self-knowledge, knowledge of donors and the coordination of fundraising efforts among the various players are requirements for success for any library fundraising effort. The engineering/science library, which plays a special supporting role for academic engineering and science departments, alumni, and the business community, can use its special strengths in collections, staff and resources to approach fundraising creatively. Those libraries that are responsive to change can improve fundraising efforts by marketing themselves as synergistic organizations open to new projects and technologies and to collaboration with a variety of partners. *[Article copies available for a fee from The Haworth Document Delivery Service: 1-800-342-9678. E-mail address: <getinfo@haworthpressinc.com> Website: <http://www.HaworthPress.com> © 2001 by The Haworth Press, Inc. All rights reserved.]*

Joanne V. Lerud, MS, MLS, is Director, Arthur Lakes Library, Colorado School of Mines, Golden, CO 80401. Lisa G. Dunn, MA, MLS, is Head of Reference, Arthur Lakes Library, Colorado School of Mines, Golden, CO 80401.

[Haworth co-indexing entry note]: "Opportunities for Creativity: Fundraising for Engineering and Science Libraries." Lerud, Joanne V., and Lisa G. Dunn. Co-published simultaneously in *Science & Technology Libraries* (The Haworth Information Press, an imprint of The Haworth Press, Inc.) Vol. 19, No. 3/4, 2001, pp. 221-235; and: *Engineering Libraries: Building Collections and Delivering Services* (ed: Thomas W. Conkling, and Linda R. Musser) The Haworth Information Press, an imprint of The Haworth Press, Inc., 2001, pp. 221-235. Single or multiple copies of this article are available for a fee from The Haworth Document Delivery Service [1-800-342-9678, 9:00 a.m. - 5:00 p.m. (EST). E-mail address: getinfo@haworthpressinc.com].

KEYWORDS. Fundraising, engineering and science libraries, academic libraries, grant writing

INTRODUCTION

Although libraries are changing in dramatic ways, certain expectations remain the same. A user expects the library to be open all hours and have all resources needed. Resources must be immediately accessible, preferably from the desktop. The challenge of supporting traditional resources as well as new technologies puts strain on already stressed acquisitions, operating and capital budgets. Engineering and science librarians, whether at a central library for a specialized institution or at a branch library for a large academic institution, are very aware of these stresses. Fundraising must be an important factor in the enhancement and even the survival of the library. Fundraising can no longer be done in one's spare time; it must be undertaken aggressively and with purpose. Grant writing is often combined with philanthropic endeavors to gain completeness of a particular effort or to get an effort off the ground. As with any funding endeavor, library fundraising seeks to find the common ground that matches the donor's wishes and the library's needs. Gifts should result in progress of the library toward its mission and goals; gifts should not force a new direction or priority. All ethical behavior incumbent in any professional interaction should be part and parcel of the fundraising endeavor. *Becoming a Fundraiser: the Principles and Practice of Library Development* (Steele and Elder 2000) is a valuable resource for the fundraiser. Its suggestions parallel practices that have been used at the Colorado School of Mines' Arthur Lakes Library.

KNOW YOUR INSTITUTION

Fundraising is done within the context of the institution, and it is imperative to know the engineering/science library, the library system and the university. The librarian as fundraiser must be able to communicate the mission and goals of the institution and must adhere to institutional requirements and policies when raising money. The engineering/science library needs a clearly articulated mission statement, goals and objectives that dovetail with those of the library system and/or university. Copies of mission statements, academic plans and annual reports from the library, engineering and science departments, main library system and university are helpful as references, as are accreditation reports, statistical compilations on library activities, and peer institution or consortium reports. Define the communities the engineering/science library serves, both institutional and public: Who are they? What are their occupations and interests? How do they use the library? Why is the library's health and growth essential to the health of the community and the culture of which it is a part?

With this information the librarian can articulate a mission and priorities for fundraising. Fundraising is an advocacy process, and the librarian should explore many ways to communicate the value of the engineering/science library to the university and to society. This communication will be accomplished differently at the university and the public levels. Although university members should see the value of the engineering/science library and may participate in fundraising for it, fundraising is usually aimed at sources outside of the university. The public generally considers libraries to be repositories of valuable information and centers of learning. In addition, members of the public or of the technical or business communities who use the engineering/science library may have very definite ideas of the library's value to themselves.

Small specialized collections within the engineering/science library may be particularly valued by individuals or organizations and may be opportunities for targeted giving. For example, the Information Center for Ropeway Studies at the Colorado School of Mines' Arthur Lakes Library, the largest collection of information on the theory, design, history, and operation of ropeway systems in the U.S., is supported on an annual basis by the International Organization for Transportation by Rope. With all specialized gifts, great caution should be exercised so that the library does not promise more than it can deliver and does not ignore its mission statement.

The library fundraiser will quickly discover that when it comes to fundraising he or she is never alone. Funding sources are finite and competition for funds within the institution is possible, so asking for funding can become a very public act. The engineering/science librarian who is part of a larger library system must follow the library system's fundraising plan. Most large library systems have long-term fundraising goals; some have completed feasibility studies to determine needs and identify donors, performed market surveys, involved a Friends group, or even appointed a development officer. If the library system has no such fundraising resources or the engineering/science librarian receives permission to pursue funding independently, the librarian must still work within the framework of university fundraising, which means working with the institutional offices responsible for fundraising and development.

THE PLAYERS

The librarian who decides to move into the arena of fundraising will find it crowded and political in nature. Fundraising on a large scale is a team effort and while everyone may not have the same priorities, they all need to agree on common fundraising goals for the library. Players include:

University Administration–The Administration must be supportive of prioritizing the library in the overall fundraising efforts of the entire institution. Without this support, the library will not be mentioned at appropriate times nor will support efforts, such as from the Development Office, be forthcoming.

Institutional Development Office–The Development Office follows the institution's development plan. Because of this role, the Development Office sometimes seems as though it is not giving the library the attention the librarian feels it deserves or is placing artificial restrictions as to whom can be contacted and when. However, Development Office staff are juggling multiple institutional needs, of which the library is only one. It is vital that fundraising reflect the library's part in a cohesive organization and not be seen as being at cross purposes with other university fundraising efforts. Development Office staff can ensure that information regarding funding priorities is properly written and distributed, assist with the proper paperwork for large gifts, and take on the responsibility of writing required reports to the donor. Depending on institutional priorities, they can also help identify prospects.

Library Director–The Library Director or Dean must be an engaged member of the fundraising team for the fundraising effort to be successful. Most donors wish to speak to the person with overall responsibility for the enterprise to see if they share a similar vision, have common philosophies or interests and are trustworthy.

Engineering/Science Library Staff–The library staff may be the first to have recognized or developed a relationship with a potential donor. Although the librarian may be prohibited by institutional fundraising policies from soliciting from donors directly to prevent conflicts with larger fundraising plans, casual conversation between the prospect and the librarian may include inquiries about funding needs. For example, although the librarians at the Arthur Lakes Library may not ask for a donation, it may be appropriate in the context of a conversation to point out a funding deficiency affecting the donor as a library patron. The engineering/science librarian with good communications skills and an awareness of the political realities of institutional fundraising can be a highly effective advocate with donors, as part of the fundraising team.

Donors–Unlike degree-granting departments, engineering/science libraries do not have alumni, a primary source of donations. However, the library does have patrons: students, engineering and science faculty, university alumni and local business people. Current and former staff members, former student assistants and volunteers are also potential donors, as are those who donate large or repeated gifts of materials (they have a pattern of giving and should not be merely categorized as "cleaning out their basement"). While adhering to the institutional fundraising plan, pay attention to those patrons who proffer the information that they are interested in financially supporting the library. A list of names from these various populations can be compiled and annotated with contact information. Donors are generally considered in two gift categories: major giving and annual giving. Major gifts are at the multi-thousand dollar level and are usually given after significant cultivation over a period of time. Annual gifts are of lesser amounts, and are generally given annually or in response to a direct appeal. Steele and Elder (2000) also define a middle range of giving of between $100-$10,000 for most libraries. These donors may be annual givers who are not able to give more, or they may be potential major donors. This is the group that should receive particular attention as these donors have already shown an interest in the library but perhaps need to find the right target to really engage their interest.

METHODS OF FUNDRAISING

The Academic Library Advancement and Development Network (ALADN) is a professional group dedicated to fundraising specifically for academic libraries. Information about what has worked in a variety of institutional environments is freely shared, whether this be campaign strategies, promotional materials, or cultivation of prospects. Additionally, this may also be an excellent venue for identifying successful fundraisers for possible recruitment. ALADN and other such groups can provide assistance and direction for fundraising endeavors. Other professional organizations such as the Association of College and Research Libraries and the Library Administration and Management Association have resources on library fundraising and promotion, such as *Library Fundraising: Models for Success* (Burlingame 1995).

Annual Appeals–One of the most common methods of fundraising for any type of library is the annual appeal. This usually consists of a letter sent to a large number of people who have some association with the library. This effort must be coordinated with the institutional Development Office so that several appeals are not sent to the same donors at the same time. The large library's Development Officer or the institutional Development Office may assist in this endeavor, and a Friends group may also be quite active. Sometimes telephone appeals accompany this effort. Annual appeals can be turned into endowments if the conditions are right. The University of Nevada's Mackay School of Mines Library targeted local companies and consultants that used the Library as a critical resource for their businesses. An endowment was established after several years of annual appeals that resulted in a sufficient donation base (Newman 1993).

Fundraising Events–Special fundraising events may be a part of the annual appeal or a part of fundraising for the library in general. Some communication with other campus offices such as the Development Office or Publicity Office should be done for the most effective results. Event planning can be very resource-intensive, especially for small libraries. Therefore, use judgement to determine what energy, time, and money will be spent and what the return on that investment might be. Fundraising events can take many forms. A gala event celebrating a major gift or purchase, such as a reception to introduce a new integrated library system, can be used to urge other gifts. Book sales are another example of a fundraising event. In this case the engineering/science library should assess the potential market for the materials it is likely to have available for sale: technical books, journals and reports. The Annual Book Sale of the Arthur Lakes Library is a multi-thousand dollar fundraiser because scientific and technical books have a wide market in the Denver metropolitan area. The staff-run Book Sale offers items weeded from the library's collections and from materials donations over the course of the year. Not only are students, faculty and alumni interested in these items, but there is a significant population of technical professionals, general members of the community and used book sellers who are avid buyers. The state requires that all are given equal access so special groups are not given previews.

Nearly any event can be an opportunity to inform people about the needs of the library and to identify prospective donors: a community art show, a book signing, an engineering design awards presentation or a retirement celebration. Alternative venues are creative ways to enlarge the database of people interested in the library and create a special connection to the library for participants, especially non-alumni. Social functions may not necessarily produce a check but might produce information that eventually brings forth a major gift. Lists of attendees, sponsors and organizers are helpful. At the Arthur Lakes Library the Director is the main lead on information gathering at these venues; staff forward any additional relevant information to the Director's office for follow-up.

Major Donors–According to Steele and Elder (2000, p. 27), "The vast majority of large gifts will come from a relatively small number of individual donors, most of whom will already be known to you or to your organization." They state that 80% of the total funds received will come from 20% of the donors. How to find those individuals? As mentioned above, the engineering/science library has no population of wealthy and successful alumni to target for major donations. However, the library should have its list of potential donors which can be evaluated to identify prospective major donors. Outside of the library, someone at an institutional level may also have identified possible prospects. If the engineering library is a priority with the university Development Office, it should be identifying prospective major donors and promoting the library as a funding opportunity.

The institutional Board of Trustees may be able to supply contacts to potential major donors–a Board member may even be a prospect if the fit is right. The Russell L. and Lyn Wood Mining History Archive of the Colorado School of Mines is one example where the interest of a Trustee meshed nicely with a stated need of the Library. The need for a specialized archive had been identified as a possible naming opportunity. During a recent capital campaign for the institution, Mr. Wood, a Trustee, and Mrs. Wood approached the Library Director with the willingness to become donors and the Archive was the perfect fit for their interests. Other university-affiliated groups can also serve as contacts. The Colorado School of Mines, like some other engineering institutions, has external visiting committees that periodically evaluate departmental programs and services. These committees are appointed by the School's Board of Trustees and are composed of industry and educational peers; the Library has its own visiting committee. While visiting committees are not fundraising bodies, in the past committee members have become so involved in a particular institutional endeavor that they became major donors in order to promote the project's success.

DEVELOPMENT

Development is a state of mind. With every patron of the engineering/science library a potential donor, the entire library staff must create an environment of advocacy for the library. Library advocacy fits seamlessly into the library's service philosophy if staff believe that what benefits the library benefits its community. However, "development" in the fundraising sense usually refers to the development and cultivation of major gifts.

Major Gifts–Solicitation of major gifts requires a significant investment of time and personnel, and the primary targets are wealthy individuals or larger corporations. Caution is indicated when selecting potential donors for major gifts. Some seem to promise or encourage continued contact but never deliver. Never accept a gift that has strings that the library may not be able commit to in the future. Gornish (1998, p. 102) provides a succinct list of the elaborate and time-consuming steps involved in the development of a major gift, which include research about the donor and repeated contacts over time. Although the Development Officer may coordinate or assist with the courting of a major gift, it is the job of the Library Director or Dean as lead person to be strongly involved and to do the actual "ask." The engineering/science librarian, if not acting as the lead person, must be ready to supply supporting information to the lead person at a moment's notice. A working relationship will develop between the Director and the prospective donor over time, including education about the needs of the Library and adjustments in expectations because of the prospective donor's feedback.

Development on the Web–Web fundraising has been used for years by a variety of organizations including libraries but is still unfamiliar territory for many academic librarians. Resources like *Fundraising and Friend-Raising on the Web* (Corson-Finnerty and Blanchard 1998) can provide guidelines. Many universities now have webpages devoted to giving. However, these pages are just as often a directory of Development Office staff or an unadorned list of previous donors, vastly under-using the Web's communication potential. Provide information relevant to potential donors as online shoppers, for example clearly defined funding opportunities they can select from, the direct benefits resulting from a gift and how donors are acknowledged. Emphasize how the gift could make a difference. The keys to Web development are: ensure that the library's fundraising link is easily identifiable from the institution's home page; incorporate a descriptive page devoted to the library; and describe specific giving opportunities. The Massachusetts Institute of Technology (MIT 2000) has created enticing development webpages that explain the importance of the library to the institution, describe its goals and direction, and list giving opportunities at various funding levels. The engineering/science library may be able to work with the engineering and science departments on campus to have a link on the departments' own webpages as well (if the groups can reconcile and coordinate their competition for funding).

Received Gifts–A significant part of development is acknowledgement of the donor. This can range from a small plaque to a listing on the library's website to a name on a room, wing or building. Naming opportunities, which are usually reserved for major gifts, can be presented to donors during the cultivation process, as part of a campaign, or as an option on a development webpage. The naming of a building is usually the top of the scale, but "named" resources can include the reference room, instruction labs, special collections, study rooms, or computer labs. The Web offers non-traditional naming opportunities (as the university permits) with public television and radio as a model, such as a database or virtual tour "supported by X Corporation."

Use the excitement from one gift to build another. Several public university libraries in Colorado have used the excitement of construction from state capital funds to ask for gifts to support and endow collections. Some of these endowed collections have been supported by large corporations moving into the geographic region who wish to have a positive community presence; some are supported by individuals who wish to commemorate their support of the institution.

GRANT WRITING

The pursuit of grants is an integral part of academia in the engineering and physical sciences, with well-practiced institutional mechanisms. Academic libraries can effectively include the pursuit of grants in their fundraising strategy, and grant writing becomes imperative if the engineering/science library is not a high priority with central library or institutional fundraising efforts. Academic libraries, however, don't always fit into the standard university grant-writing scenario. Library grants may be much smaller than the typical scientific research grant; granting agencies may be unfamiliar to administrative support staff used to working with agencies such as the U.S. Department of Energy or the National Science Foundation; justification and maintenance of funds may differ from that involving laboratory facilities, research staff and graduate students. The librarian should work with university fundraising personnel to develop appropriate strategies or go "outside the box" to find grant sources. As with all fundraising efforts, the grant proposal should reflect the engineering/science library's role within the institution. As an exploration of the library and its potential, grant seeking mirrors some activities outlined by Kirchner (1999) in his article on library advocacy when approached creatively.

Get Educated–Make full use of available expertise: institutional Development Officers, faculty from other academic departments and the network of fellow librarians. There are many "how to" books and workshops on grant writing available. Some resources are developed specifically for librarians; others are more general but still offer good advice. Copies of successful proposals as examples for the novice grant writer may be available as public documents or from colleagues. Unsuccessful proposals may be even more informative but will not be as easy to acquire. If the librarian is lucky enough to have colleagues who will share unsuccessful proposals, complete with reviewers' comments, he or she should study them carefully. (This means that sharing unsuccessful proposals should be reciprocal.) Don't turn down the opportunity to be a proposal reviewer for an internal or external committee. Being on the other side of the grant writing process can offer valuable insights into effective proposal composition.

While exploring the fundamentals of grant writing in general, become familiar with the institution's grant writing policies and procedures. It is important to be aware of any loopholes, exceptions and negotiation room in the institutional grant process. For example, the size of the grant, the purpose to which it is being put, or donor stipulations may reduce or eliminate indirect costs or overhead charges paid to the university that would otherwise take a significant portion of the award. This is particularly important for the small grants available to many libraries.

Define Projects to be Funded–Funding organizations are generally reluctant to give grants to repair overall deficiencies in the library's collections or facilities–they have no wish to serve as a substitute for adequate institutional funding. A funding organization would rather work with a partner who demonstrably values its own resources or to fund a unique or special project. Donors want to make a difference.

One of the most versatile grant writing strategies for an engineering/science library is to re-define the library's needs as an interconnected series of special projects that form a cohesive whole. Small specialized libraries have the advantage here. Alexander (1998) uses the phrase "evolving library" to refer to libraries that are small, neophyte, or developing, thereby offering opportunities for innovation, flexibility, responsiveness, synergy–characteristics that enhance their attractiveness to donors. Many libraries have special collections with access or preservation deficiencies, client populations with special needs, or staff with the expertise or enthusiasm to implement a pilot project. Even so, grant writing and project implementation resources are limited. Select as a high priority those projects that:

1. Fit into the big picture and further the library's overall mission
2. Provide staff with the expertise or resources to address multiple library goals
3. Foster partnerships leading to additional grant writing opportunities

Ideally, the project should address all three criteria.

Brainstorm with colleagues to identify funding possibilities. Browse library and higher education trade publications that contain grant awards news, such as *American Libraries, College & Research Libraries News (C&RLN)*, or the *Chronicle of Higher Education* for creative grant writing ideas. Over a three-year period, *C&RLN* included the following:

- Rutgers University: $127,502 from the New Jersey Dept. of Environmental Protection to build a New Jersey Environmental Digital Library (C&RLN 2000).
- University of Texas-Austin: $300,000 from the Welch Foundation to support statewide access to chemical information sources (C&RLN 1999).
- SUNY Buffalo: $119,000 from the NSF to a Science and Engineering Library librarian and a biology faculty member to foster case study teaching in the sciences (C&RLN 1998a).
- Navy Department Library and Naval Observatory Library: $450,000 in DOD Legacy grants to preserve and improve access to rare historic and scientific materials (C&RLN 1998b).

Above all, look at the engineering/science library from a different perspective. The primary clientele may be the engineering college or the university's physical science departments and their students, but by getting a grant to offer wireless Internet access for patrons, provide information to local industry, or partner the development of an innovative instruction program, the librarian gains grant writing experience, draws other potential donors with a success story *and* furthers the library's mission.

Formulate projects around the library's strengths to address its weaknesses. Creativity can make the ordinary funding problem into an attractive prospect for donors. Assess the library for unique collections, resources or services, and the library staff for skills and interest levels. For example, the Arthur Lakes Library was interested in digitization as an access and preservation medium but lacked expertise in this area. In collaboration with the National Mining Hall of Fame and Museum in Leadville, Colorado, a team received a grant from the Colorado Digitization Project to digitize a collection of mining photographs of historical state interest. Staff received training on technical and cataloging issues and access to project hardware and software. The result: access to the Library's collections will be improved; staff gained technical expertise and an understanding of the legal and access issues of digitization; the Library's fundraising record is enhanced; the Library becomes a good candidate for continued funding under this program; and several potential partners for future collaborative efforts have been identified.

Collaborate–Collaboration is a desirable practice for all parties in the grant process. Just like people, funding organizations want the most for their money. Giving money to a partnership can be more rewarding and reassuring than giving money to a single library or individual. Some funding organizations place a higher priority on collaborative projects, or like to see collaborative projects with specific partners such as teaching faculty, public libraries, or K-12 educational institutions. Collaboration allows the engineering/science library to take advantage of partners' expertise and resources, both in grant writing and in project implementation. For most librarians, grant writing is an activity done *in addition to* other responsibilities–the more assistance available, the better able the librarian will be to work on the funded project while juggling bibliographic instruction sessions or cataloging monographs in chemical engineering.

Reviewing the institution's missions, goals and past accomplishments as part of the initial education process of grant writing offers a good opportunity to identify any system-wide projects the library could collaborate with or piggyback on, or to identify potential partners for the library's own projects. Don't overlook faculty in other academic departments as potential partners in grant writing. The library's "information laboratory" function and the librarian's role as educator makes it a candidate for joint proposals with other science, technical or educational programs. University research centers or institutes are also potential partners if common ground can be identified.

Identify Grant Funding Sources–There are many types of grants available: government, private foundation, corporate and internal grants, each with their own demands on the grant writer. (The grant criteria don't have to say "engineering or science library" to be applicable to your library's project.) In addition to contacting the granting agency directly, grant news can be tracked by contact with local, state and regional library associations, association newsletters or the state library. Check grant and foundation reference resources such as the *Foundation Grants Index*, which lists awarded grants and provides a direct window into what foundations actually do with their money. While some resources cover a variety of grants, the *National Guide to Funding for Libraries & Information Services* focuses on foundations that have either a stated interest in or have given money specifically to libraries.

The Web is another valuable source of information. Internet resources for library grants have been around for years (Schneider 1996; Sternberg 1997; Kyker 1998). Once identified, donor webpages can provide additional information including guidelines and application forms. Check out both donor and recipient webpages for ideas. Some recipients, especially consortia or organizations with an educational component to their grant, place surprisingly detailed information about their funded projects on the Web for all to see. Donors are hoping to attract the right candidates; recipients are advertising their success, educating others, and possibly attracting partners for future projects. Both the federally funded Digital Libraries Initiative Program and the University of Illinois at Urbana-Champaign's Grainger Engineering Library (a Program participant and test site) distribute Program information via the Web (DLI 2000; UIUC 2000). The Grainger Engineering Library even includes a list of partner engineering and science publishers and information about existing and potential industrial sponsors.

Internal grants can be an overlooked source of funding. Institutions may have internal grants available for technology and classroom upgrades, distance education programs, or professional development activities. Internal grants are an excellent way for the novice to get his or her feet wet. Because evaluation takes place within the organization, the librarian may get excellent feedback, both formal and informal, on the proposal. The Arthur Lakes Library staff's first attempt at internal funding for computer equipment came back rejected but with pages of comments from the review committee and informal suggestions from a reviewer. Their second attempt and a later proposal were both successful. Internal grants also usually have more informal submission criteria that take less time and effort to fulfill. The rule here is that no grant is too small if the effort is correspondingly minimal. An hour's worth of time spent reading the proposal guidelines for an internal classroom upgrade grant and writing the proposal got the Arthur Lakes Library fittings for some of its study rooms and a projection screen for presentations.

When identifying and selecting any granting organization, pay close attention to the information the donors themselves supply about their mission and stated goals, as well as their history of giving. Most will be very explicit about why and to whom they give money. The project needs must be matched with the donor's desires–don't waste time soliciting money from organizations that clearly have no interest in giving to an unrelated project.

Write the Proposal–The donor ultimately wants the library as much as the library wants their money. "Think of a relationship with a foundation as a partnership. They have the money. You have the ideas, the service mission, and the human resources to actualize their philanthropic impulses, giving them concrete form" (Lee and Hunt 1992, p.37). Grant proposals just need to help donors make up their minds.

As with all fundraising, the proposal should reflect both organizations' missions and emphasize those areas where they dovetail. Many proposals are rejected because the proposal writers failed to identify their audience or fell in love with their project to the extent that they forgot that they were not the only players in this game–the donor is the one with the funds and they need a good reason to buy in to the project. Line up matching funds or shared costs. Just as donors like to see the recipient collaborate with others, they like to feel that they are a part of a collaboration. Make the donor feel like a partner with both the library and other contributors. Many donors ask for a description of matching funds, in-kind contributions, or other ways the cost of the project can be shared. Communicate with the donor if you have a contribution other than those acceptable contributions outlined in the proposal guidelines.

There are basic technical communication issues that must be addressed in the proposal writing stage. The most obvious: follow the directions. Many proposals are rejected because the proposal writer simply didn't follow the directions. Like a job application, read the application criteria carefully and provide all requested information in an acceptable format. Get the most current submission guidelines, preferably by contacting the agency directly. Direct communication with donors, if acceptable to them, can be very effective. Ask for advice, seek clarification, even ask how much money you should ask for. Again, as in applying for a job, clear up any possible ambiguities in this way rather than requiring that the funding agencies do so. Even if extensive direct communication with the donor takes place, put care into the written proposal. Communicate clearly. Eliminate jargon that may be indecipherable to the donor's review committee and try to target that specific audience. A technology upgrade proposal from the Arthur Lakes Library was returned for editing because it provided too many technical details–the donors wanted to know what the Library was doing, not how it was being done. A different technology upgrade proposal sent to another group was returned for editing because there were not enough technical details–the reviewers in this case included computer support personnel who wanted that information.

Success and Rejection–Successful grant proposals encourage further success, but rejection is inevitable when donors receive thousands of proposals each year. Expect to face rejection some of the time even when planning for success. It is more fulfilling to approach grant writing, as in all fundraising, as advocacy for the library. Impress the granting agency; make them remember the library; treat reviewers like skeptical patrons and win them over. Repeat the grant writing process until success is achieved. Rejections themselves can be an excellent learning exercise. Listen carefully to every "no" and follow up on any and every suggestion by reviewers. The slightest hint that the proposal be re-submitted should be acted upon–remember, the donor wants the library to succeed.

CONCLUSION

Fundraising is easiest when the planning has been done in advance. Even if the engineering/science librarian doesn't have institutional permission to fundraise or write grants, he or she may still be unexpectedly presented with a funding opportunity if the proper justification can be produced in time. With thought and involvement invested in the fundraising process, the librarian will be better able to respond to both planned and unplanned funding opportunities. Compile lists of potential donors. Involve the staff in fundraising plans. Maintain a file of institutional information that can be referenced to compose summaries and fact sheets; network with colleagues on possible collaborative efforts and let them know that the library is a potential partner. If the university doesn't permit library staff to fundraise directly, prepare fundraising plans for internal reference. Keep draft proposals ready to modify when funding opportunities arise; have a "working document" ready to supply to the person who is assigned fundraising responsibilities–this is especially useful if this person is not a librarian and needs supporting documentation to effectively fundraise. Above all, get involved in the process of fundraising. There is money available to creative engineering and science libraries–just ask.

REFERENCES

Alexander, Johanna Olson. 1998. Fundraising for the evolving academic library: the strategic small shop advantage. Journal of Academic Librarianship 24(2): 131-138.
Burlingame, Dwight F., ed. 1995. Library Fundraising: Models for Success. Chicago IL: American Library Association.
C&RLN. 2000. Grants and acquisitions. C&RL News 61(4): 326.

C&RLN. 1999. Grants and acquisitions. C&RL News 60(11): 963.

C&RLN. 1998a. Grants and acquisitions. C&RL News 59(10): 795-796.

C&RLN. 1998b. Grants and acquisitions. C&RL News 59(6): 456.

Corson-Finnerty, Adam; Blanchard, Laura. 1998. Fundraising and Friend-raising on the Web. Chicago IL: American Library Association.

DLI. 2000. Digital Libraries Initiative: A community of researchers and agencies working together to bring the world's knowledge to your desktop, http://www.dli2.nsf.gov/. Washington DC: National Science Foundation, accessed 9/27/2000.

Gornish, Stanley E. 1998. How to apply fund-raising principles in a competitive environment. Library Administration & Management 12(2): 94-103.

Kirchner, Terry. 1999. Advocacy 101 for academic libraries. C&RL News 60(10): 844-846, 849.

Kyker, Penny. 1998. Selected World Wide Web sites for library grants and fund-raising. Library Administration & Management 12(2): 64-71.

Lee, Hwa-wei; Hunt, Gary A. 1992. Fundraising for the 1990s: the Challenge Ahead. A Practical Guide for Library Fundraising from Novice to Expert. Canfield OH: Genaway.

MIT. 2000. The campaign for MIT: The Libraries, http://web.mit.edu/campaign/priorities/library3.html. Cambridge MA: Massachusetts Institute of Technology, accessed 10/20/2000.

Newman, Linda. 1993. The Mines Library endowment, a ten-year report. In: Geoscience Information Society Twenty-eighth Meeting Proceedings, Boston MA Oct. 25-28 1993 v. 24., Wick, Connie, ed. Alexandria VA: Geoscience Information Society, 117-119.

Schneider, Karen G. 1996. Grabbing grants via the Internet. American Libraries 27(3): 66-67.

Steele, Victoria; Elder, Stephen D. 2000. Becoming a Fundraiser: the Principles and Practice of Library Development. 2nd ed. Chicago IL: American Library Association.

Sternberg, Hilary. 1997. Internet resources for grants and foundations. College and Research Libraries News 58(5): 314.

UIUC. 2000. University of Illinois at Urbana-Champaign Digital Libraries Initiative, http://www.dli.grainger.uiuc.edu/. Champaign IL: University of Illinois at Urbana-Champaign, accessed 12/20/2000.

Index

Page numbers followed by *f* indicate figures; those followed by *t* indicate tables; and those followed by *n* indicate notes.

Printed in the United States
by Baker & Taylor Publisher Services